Fungal Diversity in the Mediterranean Area

Fungal Diversity in the Mediterranean Area

Editor

Giuseppe Venturella

MDPI • Basel • Beijing • Wuhan • Barcelona • Belgrade • Manchester • Tokyo • Cluj • Tianjin

Editor
Giuseppe Venturella
University of Palermo
Italy

Editorial Office
MDPI
St. Alban-Anlage 66
4052 Basel, Switzerland

This is a reprint of articles from the Special Issue published online in the open access journal *Diversity* (ISSN 1424-2818) (available at: https://www.mdpi.com/journal/diversity/special_issues/fungal_diversity).

For citation purposes, cite each article independently as indicated on the article page online and as indicated below:

LastName, A.A.; LastName, B.B.; LastName, C.C. Article Title. *Journal Name* **Year**, *Article Number*, Page Range.

ISBN 978-3-03936-978-2 (Hbk)
ISBN 978-3-03936-979-9 (PDF)

Contents

About the Editor

Giuseppe Venturella, Prof., Full Professor of Forest Botany and Mycology at the Department of Agricultural, Food and Forestry Sciences (SAAF) of the University of Palermo, he is also the President of the Italian Medicinal Mushrooms Society (SIFM) and a member of the Mycology Interest Group of the Italian Botanical Society (SBI). He is an Italian representative in the International Society of Medicinal Mushrooms, a member of the Editorial Board of the *International Journal of Medicinal Mushrooms*, as well as the President of the Ninth International Medicinal Mushrooms Conference, held in Palermo in 2017. Finally, he is the author of 108 scientific publications indexed on SCOPUS, and numerous other publications in national and international journals, as well as monographs on mushrooms and truffles.

diversity

MDPI

Editorial

Fungal Diversity in the Mediterranean Area

Giuseppe Venturella

Department of Agricultural, Food and Forest Sciences, University of Palermo, Viale delle Scienze, Bldg. 5, I-90128 Palermo, Italy; giuseppe.venturella@unipa.it; Tel.: +39-09123891234

Received: 19 June 2020; Accepted: 19 June 2020; Published: 21 June 2020

Abstract: The Special Issue entitled "Fungal Diversity in the Mediterranean Area" aimed at highlighting the role of various organisms in the Mediterranean habitat. The role of fungi at the root and phyllosphere level; the biodiversity in small island territories and the sea; rare forms of fungi never previously found; the commercial, food, and therapeutic value of some ascomycetes and basidiomycetes; the diversity related to fungi associated with galls on plants; and the important role of culture collection for the ex situ conservation of fungal biodiversity are the topics dealt with in this Special Issue.

Keywords: fungal diversity; mycorrhiza; Mediterranean forest; medicinal mushroom; bioprospecting; marine fungi; phylogenetics; galls; basidiomycetes; ascomycetes; culture collection

---

Fungi are extremely heterogeneous organisms characterized by high levels of species diversity and are widespread in all environments. Research on fungal diversity cannot be considered exhaustive, given the continuous discovery of new species and the variability of environments where fungi can be harvested, including the seabed. The fields of application are also varied and range from agriculture, forestry, food, medical, and pharmaceutical sectors. If compared to the central and northern European regions, the Mediterranean environment is a reservoir of continuous discoveries which, in addition to having a taxonomic, environmental, and biogeographical interest, allow researchers to highlight peculiar contents of nutritive elements and uncommon therapeutic applications. This Special Issue includes eight research articles dealing with the fungal biodiversity of the Mediterranean area from various points of view.

Mahmoudi et al. compare samples of roots and rhizospheric soils from arid areas of Tunisia characterized by intensive grazing [1]. The mycorrhizal frequency and the intensity and density of spores varies between plants at the same site and, for each plant, between sites.; Mahmoudi et al. have shown a positive effect of mycorrhizal plants on the microbial activity of the soil. The authors conclude that Arbuscular Mycorrhizal Fungi (AMF) improves soil biological properties, supporting the hypothesis that mycorrhiza and grazing compete for plant photosynthates. Besides, under arid conditions, mycorrhizal symbiosis plays a decisive role concerning soil functionality.

The importance of mycorrhizae is even more evident in the case of species of high historical, gastronomic, and commercial value. *Tuber magnatum* Pico, the most prized truffle in the world, has been studied by Belfiori et al. who examined white truffles from Italy, Hungary, Serbia, Romania, Bulgaria, and Greece and characterized them from a genetic point of view. This study is of fundamental importance for application purposes and to allow the better traceability of white truffles for commercial use and also to prevent the erosion of the biodiversity of white truffles [2].

The biodiversity of macromycetes in Mediterranean forests is the theme of the scientific contributions of Polemis et al. and Gargano et al. In the first article, the authors analyze the fungal diversity of the basidiomycetes associated with *Alnus glutinosa* L. in a restricted environment such as the island of Andros in the Cyclades (Greece). In a long term study, the authors analyze from a morphological, ecological and genetic point of view several macromycetes, of which 21 species are first national records and 68 are reported for the first time from Greek *Alnus glutinosa* forests, including some rare species [3].

Gargano et al. investigated a rare species of albino maitake (*Grifola frondosa* (Dicks.) Gray) collected for the first time in a forest ecosystem of Sicily (southern Italy) [4]. The article highlights the potential application of the albino maitake concerning its nutritional value, particularly high in certain mineral elements and vitamins, and medical value about the ability of its extracts to reduce the production of biofilm by *Staphylococcus aureus* ATCC 43300.

Lazarević and Menkis also highlight how the phyllosphere is expressive of high species diversity. In the case study of the phyllosphere of the endemic forest tree *Pinus heldreichii* H.Christ., a huge number of fungal species were isolated, and mainly constituted Ascomycota [5]. The variability of the fungal community detected at different study sites and altitudes highlights the influence of environmental conditions on the presence/absence of fungal species. There is also a significant correlation between the presence of pathogenic fungi on the leaves, exalted by biotic and abiotic stress factors, and the composition of the fungal community.

The Special Issue also includes an investigation into the diversity of marine fungi by Poli et al. These authors reported the presence of new genera and species isolated from seagrass and algae of the Mediterranean Sea and highlighted how the families Roussoellaceae and Thyridariaceae, until now associated with terrestrial plants, are well represented also in the marine environment [6].

Zimowska et al. contributed to a particular aspect of fungal diversity related to fungi associated with galls on plants of the family Lamiaceae. The results showed full identity with *Botryosphaeria dothidea* (Moug.) Ces. & De Not. of isolates from galls collected from Lamiaceae, while a possible separation from this species should be verified for isolates recovered from *Acacia* in Australia and South Africa [7].

Finally, an interesting contribution to the ex situ conservation of wood decay fungi has been published by Girometta et al. The strains, kept in the MicUNIPV Research Culture Collection of the University of Pavia (Italy), include some species of environmental and medicinal interest closely related to the Mediterranean environment sensu stricto, together with others typical of environments characterized by continental temperate climates [8].

The articles published in this Special Issue reaffirm the importance and role of fungi in different ecosystems. The characterization of fungal biodiversity is of fundamental importance both from an environmental and applicative point of view. Further studies should be conducted in the future to highlight the importance of the in situ and ex situ conservation of fungal diversity for future generations.

Funding: This research received no external funding.

Conflicts of Interest: The author declares no conflict of interest.

References

1. Mahmoudi, N.; Dias, T.; Mahdhi, M.; Cruz, C.; Mars, M.; Caeiro, M.F. Does Arbuscular Mycorrhiza Determine Soil Microbial Functionality in Nutrient-Limited Mediterranean Arid Ecosystems? *Diversity* **2020**, *12*, 234. [CrossRef]
2. Belfiori, B.; D'Angelo, V.; Riccioni, C.; Leonardi, M.; Paolocci, F.; Pacioni, G.; Rubini, A. Genetic Structure and Phylogeography of *Tuber magnatum* Populations. *Diversity* **2020**, *12*, 44. [CrossRef]
3. Polemis, E.; Fryssouli, V.; Daskalopoulos, V.; Zervakis, G.I. Basidiomycetes Associated with *Alnus glutinosa* Habitats in Andros Island (Cyclades, Greece). *Diversity* **2020**, *12*, 232. [CrossRef]
4. Gargano, M.L.; Zervakis, G.I.; Isikhuemhen, O.S.; Venturella, G.; Calvo, R.; Giammanco, A.; Fasciana, T.; Ferraro, V. Ecology, Phylogeny, and Potential Nutritional and Medicinal Value of a Rare White "Maitake" Collected in a Mediterranean Forest. *Diversity* **2020**, *12*, 230. [CrossRef]
5. Lazarević, J.; Menkis, A. Fungal Diversity in the Phyllosphere of *Pinus heldreichii* H. Christ—An Endemic and High-Altitude Pine of the Mediterranean Region. *Diversity* **2020**, *12*, 172. [CrossRef]
6. Poli, A.; Bovio, E.; Ranieri, L.; Varese, G.C.; Prigione, V. News from the Sea: A New Genus and Seven New Species in the Pleosporalean Families Roussoellaceae and Thyridariaceae. *Diversity* **2020**, *12*, 144. [CrossRef]
7. Zimowska, B.; Oko, S.; Becchimanzi, A.; Krol, E.D.; Nicoletti, R. Phylogenetic Characterization of *Botryosphaeria* Strains Associated with *Asphondylia* Galls on Species of Lamiaceae. *Diversity* **2020**, *12*, 41. [CrossRef]

8. Girometta, C.E.; Bernicchia, A.; Baiguera, R.M.; Bracco, F.; Buratti, S.; Cartabia, M.; Picco, A.M.; Savino, E. An Italian Research Culture Collection of Wood Decay Fungi. *Diversity* **2020**, *12*, 58. [CrossRef]

diversity

MDPI

Article

Basidiomycetes Associated with *Alnus glutinosa* Habitats in Andros Island (Cyclades, Greece)

Elias Polemis, Vassiliki Fryssouli, Vassileios Daskalopoulos and Georgios I. Zervakis *

Laboratory of General and Agricultural Microbiology, Agricultural University of Athens, 11855 Athens, Greece; teonanac_rec1@hotmail.com (E.P.); vfrisouli@gmail.com (V.F.); vassilismks@gmail.com (V.D.)

* Correspondence: zervakis@aua.gr; Tel.: +30-210-5294341

Received: 15 May 2020; Accepted: 7 June 2020; Published: 9 June 2020

Abstract: Alluvial forests dominated by black alder (*Alnus glutinosa*) are widespread in Europe along river banks and watercourses forming a habitat of renowned ecological/conservation importance. Despite the considerable interest this habitat has attracted in terms of the associated fungal diversity, very few pertinent data are available from the eastern Mediterranean. Andros island (Aegean Sea, Greece) hosts the southernmost population of *A. glutinosa* in the Balkan Peninsula; such stands have been systematically inventoried for several years in respect to macrofungi. In total, 187 specimens were collected and studied by examining morphoanatomic features and by evaluating (when necessary) the outcome of sequencing the internal transcribed spacer (ITS) region of nuclear ribosomal DNA (nrDNA) to elucidate their identity and obtain an insight into phylogenetic relationships. As a result, 106 species were recorded, 92 are saprotrophic and 14 form ectomycorrhizae (ECM) with alders. Twenty-one species are first national records, while 68 other species are reported for the first time from this habitat in Greece. Several findings of particular interest due to their rarity, ecological preferences and/or taxonomic status are presented in detail and discussed, e.g., six *Alnicola* taxa, *Cortinarius americanus*, *Lactarius obscuratus*, *Paxillus olivellus* and *Russula pumila* (among the ECMs), and the saprotrophs *Entoloma uranochroum*, *Gymnopilus arenophilus*, *Hyphoderma nemorale*, *Lepiota ochraceofulva*, *Phanerochaete livescens* and *Psathyrella hellebosensis*.

Keywords: macrofungi; Basidiomycota; mushroom diversity; ectomycorrhiza; saprotroph; alder; Aegean Sea; Mediterranean; *Alnicola*

1. Introduction

Alluvial forests with *Alnus glutinosa* Gaertn. and *Fraxinus excelsior* L. (priority habitat 91E0*; Annex I, Directive 92/43/EEC) are distributed throughout Europe, but they are generally rare and threatened since only remnants exist, mainly in central and northern Europe [1]. Alder stands are considerably less frequent in the Mediterranean region, where the repercussions of changes in the hydrological cycle caused by global warming and climate destabilization are much more evident [2]. The southernmost limit of the priority habitat 91E0* in the Balkan Peninsula is located in Andros island (Figure 1), i.e., the northernmost in the Cyclades and situated at a transition zone between continental Greece and other islands of the Aegean Archipelago. From the geomorphological point of view, it is characterized by a remarkably intense relief and by many rivulets and streams of constant flow, which are unique among most of Central and South Aegean islands. *A. glutinosa* trees demonstrate a patchy distribution in Andros, predominantly occurring along the main streams within the Site of Community Importance (SCI) GR4220001 and in altitudes ranging from sea level to as high as 850 m above sea level (a.s.l.), very close to the highest peaks of the island. In many cases black alders are mixed with *Platanus orientalis* L., *Fraxinus ornus* L. and/or *Nerium oleander* L. (in lower altitudes), while they also form pure stands, as it is the case at the estuaries of the Vori stream in NE Andros.

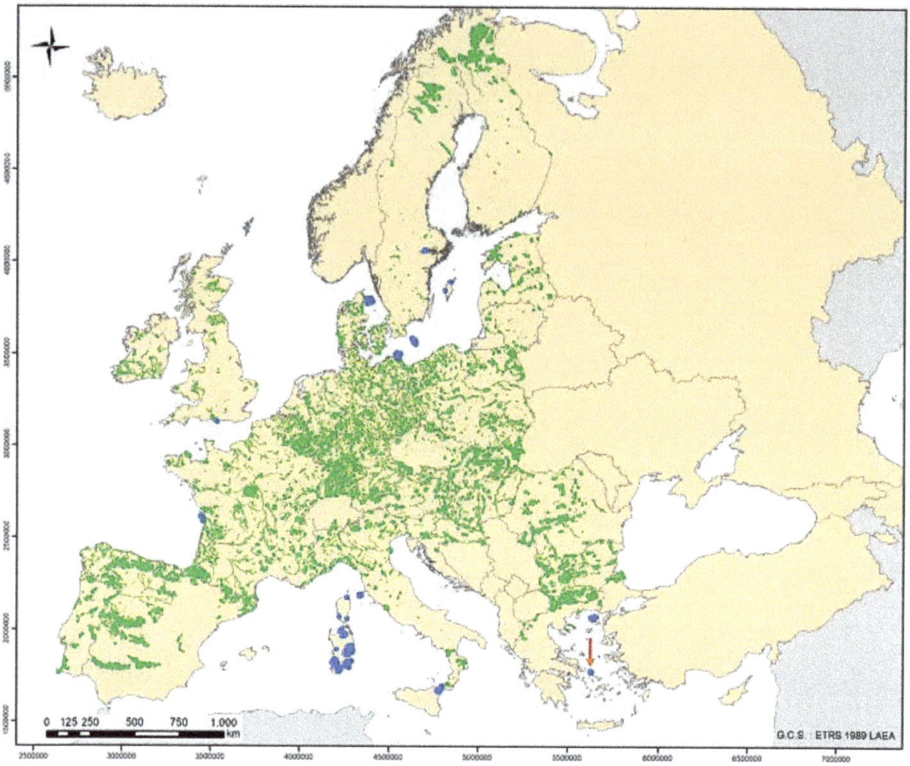

Figure 1. Map presenting Natura 2000 sites, which include the priority habitat 91E0* in continental Europe (in green) and in islands (in blue); Andros island is indicated by the red arrow. Data from https://www.eea.europa.eu/data-and-maps/data/natura-6.

Alder trees are known to form symbiotic relationships with nitrogen-fixing actinomycetes of the genus *Frankia* Brunchorst [3,4], with arbuscular mycorrhizal fungi (AM) of Glomeromycota [5,6] and with various ectomycorrhizal (ECM) fungi of Ascomycota and Basidiomycota [7–9]. European alder stands have been relatively well-studied in terms of both macro- and microfungal communities, and approx. 1000 species of saprotrophic and ECM macrofungi were reported [10–15]. In addition, mycocoenological studies from Europe and North America suggested that ECM fungi of *Alnus* spp. exhibit a remarkably high degree of host specificity compared to other tree species [8,16], while the analysis of both sporophores and ectomycorrhizae evidenced that alders have a low number (<50) of ECM symbionts worldwide [17–19].

Limited knowledge is available on the diversity of fungi associated with alders in Greece, and only preliminary data are reported in the few pertinent publications [20,21]. On the other hand, Andros is the only island of the Aegean Archipelago where a systematic inventory of macrofungi is in progress for more than 20 years. Biotopes characterized by river banks, springs and alluvial forests, where *A. glutinosa* is often the dominant tree species, were forayed in the past and 37 mushroom species were reported from this particular habitat in Andros, including ECM symbionts as well as xylotrophic, litter and/or humus saprotrophs [22–24]. Among the latter, *Entoloma alnicola* Noordel. & Polemis was described asnew species for science and it is still known from the type locality only [25].

Since 2017, mycodiversity studies in alder stands of Andros were intensified in the frame of a LIFE Nature project (LIFE16-NAT_GR_000606), which -among others- aims at the conservation and restoration of the priority habitat 91E0* in the island. Hence, during the last few years, new sites with alder stands were repeatedly forayed (in addition to those previously investigated), and a large number of new collections were made. These, together with previously sampled—but still unidentified—specimens, were subjected to detailed morphoanatomical examination in conjunction with sequencing and phylogenetic analyses (where judged necessary) in order to assess their identity. Moreover, in several occasions, past relevant reports on recorded taxa were revised/re-evaluated according to the latest respective taxonomic and phylogenetic concepts. Hence, this work presents an updated compilation of available data on the diversity of macrofungi in a habitat of significant interest occurring at the limits of its distribution in Europe.

2. Materials and Methods

2.1. *Sampling of Biological Material*

Data presented in this inventory are based on specimens collected from 10 sampling sites covering almost the entire area of *A. glutinosa* distribution in Andros island, which appears mainly within (or marginally out) the SCI GR4220001, extending from sea-level to an altitude of ca. 850 m a.s.l. (Figure 2; Table S1).The biological material examined for the purpose of this work was sampled in 38 forays performed during the last 25 years from late October to April; more than half of those (#23) were conducted in the period from 2017 to 2020. In total, 187 specimens found exclusively under alder trees or directly on their wood, woody residues or leaf-litter were collected, and voucher specimens are deposited in the Fungarium of the Laboratory of General and Agricultural University of Athens (ACAM).

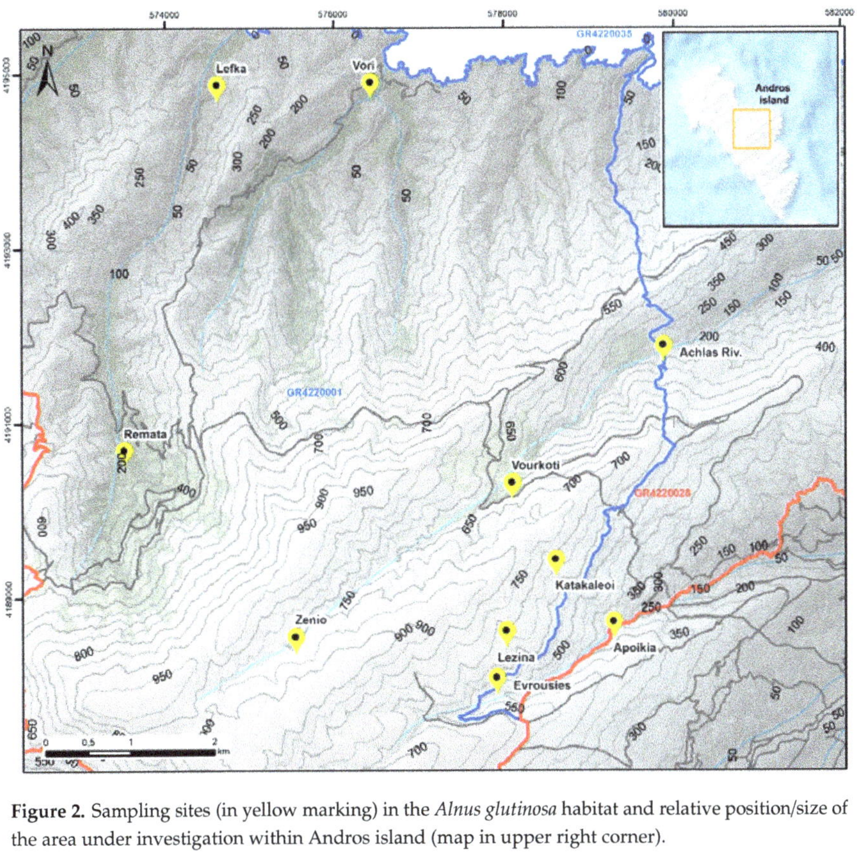

Figure 2. Sampling sites (in yellow marking) in the *Alnus glutinosa* habitat and relative position/size of the area under investigation within Andros island (map in upper right corner).

2.2. Morpho-Anatomical Features in Basidiomes

The morphological study included in situ recording of macroscopic features of taxonomic interest, while ex-situ examination involved observations of morphoanatomical characters in dried specimens. Sections were mounted and observed in KOH 3–5% (*w/v*), in Melzer's reagent, in cotton-blue, in cresyl-blue and in sulfovaniline solution. Observations were performed with the use of a Zeiss AxioImager A2 microscope under bright field and differential interference contrast (DIC); microphotographs were taken with the aid of a mounted digital camera (Axiocam). For all examined specimens a minimum of 30 mature basidiospores were measured and the resulting measurements as well as additional observations of other essential microscopical features (hymenial cystidia, pileipellis etc.) were used for determination of the species examined in accordance to pertinent identification keys and monographs (e.g., [26–36]).

2.3. DNA Extraction, Amplification and Sequencing

When deemed necessary, DNA sequencing and phylogenetic analyses were performed. Total genomic DNA was obtained from dried basidiomes and DNA extraction was performed through the use of the Nucleospin Plant II DNA kit (Macherey and Nagel, Düren, Germany) by following the manufacturer's protocol. The internal transcribed spacer (ITS; ITS1, 5.8S, ITS2) region within the nuclear ribosomal RNA gene cluster was examined by using the primers ITS1/ITS4 [37]. Polymerase chain reactions (PCR) were performed in 50 μL containing 50 ng DNA template, 0.25 μM of each primer, 0.2 mM of each dNTP, 1 × HiFi Buffer (Takara BIO INC., Shiga, Japan) and 1 U HiFi Taq DNA polymerase (Takara BIO INC., Shiga, Japan). PCR reactions were performed as follows: 94 °C for 5 min, followed by 35 cycles of 94 °C for 30 s, 50 °C for 30 s and 72 °C for 1 min, and a final extension at 72 °C for 10 min. PCR products were run in 1% agarose gels and purified using Invitrogen PureLink kit (Thermo Fisher Scientific, Seoul, S. Korea), and were submitted for sequencing to CeMIA SA (Larissa, Greece). The same PCR primers were used for sequencing. Chromatograms were checked with the aid of BioEdit v. 7.2.5 software [38]. Then sequences were examined against GenBank built-in search tools for obtaining information which could confer at identifying the material under study. A total of 61 validated sequences generated in this work were deposited in GenBank and the accession numbers MT458502 to MT458562 were obtained.

2.4. Phylogenetic Analysis of Sequence Data

A total of 42, 29 and 22 ITS sequences corresponding to selected species of the genera *Alnicola* Kühner (and *Naucoria* (Fr.) P. Kumm.), *Lactarius* Pers. and *Paxillus* Fr. (including 12, 5 and 4 sequences generated in this work), respectively, were subjected to phylogenetic analysis. In addition, species of the same or other genera were used as outgroups in each case. Multiple sequence alignment of each ITS rDNA dataset was conducted using the Q-INS-I algorithm as implemented in the online version of MAFFT v. 7 [39]. Alignments were reviewed, manually adjusted at misaligned sites and trimmed at the same position through MEGA X [40] before being used for further analysis.

Phylogenetic relationships of taxa for each alignment were inferred by using maximum likelihood (ML) and Bayesian inference (BI) through the CIPRES web portal (www.phylo.org; Miller et al. 2010). ML analyses were conducted by RAxML BlackBox online server (http://phylobench.vital-it.ch/raxml-bb/) [41] using default parameters and calculating bootstrap statistics according to the program recommendations for the best-scoring ML tree. BI analyses were performed by MrBayes v. 3.2.1 [42]. The best-fit substitution model for each dataset was selected according to the corrected Akaike information criterion (cAIC), as implemented in jModeltest v.2 [43]. The TPM2uf+G, TPM1uf+G and SYM+G models were selected for the *Alnicola*, *Lactarius* and *Paxillus* datasets, respectively. To estimate posterior probabilities, Markov chain Monte Carlo (MCMC) simulation was implemented in two parallel independent runs of four chains, one cold and three heated, with trees sampled every 1000 generations until the standard deviation of split frequencies is below 0.05; the first 25% of trees were omitted as burn-in. A 50% majority rule consensus tree was built and visualized with iTOL [44]. Clades with ML bootstrap support (MLB) ≥ 65% and Bayesian posterior probability (BPP) ≥ 95% were considered as significantly supported.

3. Results and Discussion

The study of 187 specimens of macrofungi associated with the *A. glutinosa* priority habitat in Andros led to the identification of 106 species (74 genera) of basidiomycetes. Among them, 14 (13%) are ECM species (Table 1) strictly associated with alders [18,19]. The other 92 (87%) are saprotrophic; 70 (66%) saproxylic and 22 (21%) saprotrophic on soil, humus or leaf-litter (Table 2). Interestingly, 10 ECM and 11 saprotrophic species are first national records, while other 68 are reported for the first time from this habitat in Greece. Identification of specimens to species was performed by examining their morphoanatomic features and by evaluating (when necessary) the outcome of ITS sequencing and phylogenetic analysis; in the latter case, the respective GenBank accession numbers are provided (Tables 1 and 2). Selected findings of particular interest are presented (and discussed) by providing brief descriptions and comments on characters of potentially diagnostic value.

Table 1. Ectomycorrhizal (ECM) fungi identified during the study: species name, specimen code/collection date, locality and GenBank accession numbers for ITS sequences generated. First national records for Greece are indicated by an asterisk (*) before the species name.

a/a	Species Name	Specimen Code/ Collection Date	Locality	GenBank Accession No.
1	**Alnicola escharoides* (Fr.) Romagn.	EP.17-A1344/11-Nov-2017 EP.17-A1420/24-Nov-2017 EP.18-A1548/22-Feb-2018 EP.18-A1561/1-Nov-2018 EP.18-A1571/2-Nov-2018 EP.19-A1636/16 Nov 2019	Zenio Vori Vori Zenio Vourkoti Katakalaioi	 MT458538 MT458539 MT458540
2	**Alnicola inculta* (Peck) Singer	EP.17-A1346/11 Nov 2017	Zenio	MT458541
3	**Alnicola luteolofibrillosa* Kühner	EP.17-A1430/24-Nov-2017	Vori	MT458542
4	**Alnicola subconspersa* (Kühner ex P.D. Orton) Bon	EP.17-A1421/24-Nov-2017 EP.19-A1637/16-Nov-2019	Vori Katakalaioi	MT458543 MT458544
5	*Alnicola striatula* (P.D. Orton) Romagn.	EP.04-A679/15-Nov-2004 EP.19-A1614/14-Nov-2019	Evrousies Evrousies	 MT458545
6	**Alnicola umbrina* (R. Maire) Kühner	EP.04-A678/15-Nov-2004 EP.17-A1377/2-Nov-2017 EP.18-A1572/2-Nov-2018 EP.19-A1607/12 Nov 2019 EP.19-A1638/16-Nov-2019 EP.19-A1646/17-Nov-2019 EP.19-A1666/2-Dec-2019	Evrousies Lezina Vourkoti Zenio Katakalaioi Achlas riv. Remata	 MT458546 MT458547 MT458548 MT458549
7	**Cortinarius americanus* A.H. Sm.	EP.19-A1622/15-Nov-2019	Vourkoti	
8	*Gyrodon lividus* (Bull.) Sacc.	EP.14-A1263/1-Nov-2014 EP.17-A1428/24-Nov-2017	Vori Vori	
9	**Inocybe calospora* Quél.	EP.18-A1570/2-Nov-2018	Vourkoti	MT458550
10	**Lactarius obscuratus* (Lasch) Fr.	EP.17-A1347/11-Oct-2017 EP.17-A1566/1-Nov-2018 EP.17-A1576/2-Nov-2018 EP.19-A1645/17-Nov-2019 EP.19-A1664/30-Nov-2019	Zenio Zenio Vourkoti Achlas riv. Remata	MT458551 MT458552 MT458553 MT458554 MT458555

Table 1. *Cont.*

a/a	Species Name	Specimen Code/ Collection Date	Locality	GenBank Accession No.
11	**Paxillus olivellus* P.-A. Moreau, J.-P. Chaumeton, Gryta & Jarge	EP.95-A028/13-Nov-1995 EP.02-A353/22-Sep-2002 EP.04-A670/23-Oct-2004 EP.04-A673/24-Oct-2004 EP.14-A1266/1-Nov-2014 EP.17-A1348/11-Nov-2017 EP.17-A1396/23-Nov-2017 EP.17-A1426/24-Nov-2017 EP.18-A1552/22-Feb-2018 EP.18-A1583/2-Nov-2018 EP.19-A1628/16-Nov-2019	Achlas riv. Evrousies Remata Achlas riv. Vori Zenio Evrousies Vori Vori Vourkoti Katakalaioi	 MT458556 MT458557 MT458558 MT458559
12	**Russula pumila* Rouzeau & F. Massart	EP.18-A1575/2-Nov-2018	Vourkoti	MT458560
13	*Tomentella stuposa* (Link) Stalpers	EP.02-A327/29-Apr-2002	Vori	MT458561
14	*Tomentella sublilacina* (Ellis & Holw.) Wakef.	EP.02-A452/11-Oct-2002 EP.17-A1437/24-Nov-2017	Achlas riv. Vori	 MT458562

Table 2. Saprotrophic basidiomycetes identified during the study: species name, specimen code/collection date, locality, type of substrate and GenBank accession numbers for ITS sequences generated. First national records for Greece are indicated by an asterisk (*) before the species name.

a/a	Species Name	Specimen Code/Collection Date	Locality	Substrate Type	GenBank Accession No.
1	*Abortiporus biennis* (Bull.) Singer	EP.18-A1582/02-Nov-2018	Vourkoti	fallen trunk	
2	*Agaricus moelleri* Wasser	EP.19-A1613/14-Nov-2019	Katakalaioi	leaf-litter	MT458502
3	*Amaropostia stiptica* (Pers.) B.K. Cui, L.L. Shen & Y.C. Dai	EP.17-A1423/24-Nov-2017	Vori	dead stump	MT458503
4	*Armillaria gallica* Marxm. & Romagn.	EP.17-A1443/24-Nov-2017	Vori	dead stump	
5	*Armillaria mellea* (Vahl) P. Kumm.	EP.95-A021/12-Nov-1995 EP.18-A1584/02-Nov-2018 EP.19-A1651/29-Nov-2019	Remata Vourkoti Vori	dead stump standing trunk trunk base	
6	*Auricularia auricula-judae* (Bull.) Quél.	EP.18-A1539/22-Feb-2018 EP.19-A1672/02-Dec-2019	Vori Remata	standing trunk standing trunk	
7	*Bjerkandera adusta* (Willd.) P. Karst.	EP.19-A1670/02-Dec-2019	Remata	fallen trunk	
8	*Botryobasidium candicans* J. Erikss.	EP.01-A275/26-Dec-2001 EP.11-A1023/05-Jan-2011 EP.17-A1434/24-Nov-2017	Vori Vori Vori	fallen trunk fallen trunk fallen trunk	
9	*Brevicellicium olivascens* (Bres.) K.H. Larss. & Hjortstam	EP.17-A1352/11-Nov-2017	Zenio	fallen trunk	MT458504
10	*Calocera cornea* (Batsch) Fr.	EP.17-A1440/25-Nov-2017 EP.19-A1616/14-Nov-2019	Vori Evrousies	dead stump dead stump	
11	*Ceriporia purpurea* (Fr.) Donk	EP.17-A1350/11-Nov-2017 EP.17-A1363/11-Nov-2017	Zenio Zenio	fallen trunk fallen trunk	MT458505
12	*Chondrostereum purpureum* (Pers.) Pouzar	EP.17-A1467/28-Nov-2017 EP.19-A1678/02-Dec-2019	Achlas riv. Remata	standing trunk standing trunk	MT458506
13	*Clavaria fragilis* Holmsk.	EP.19-A1639/16-Nov-2019	Katakalaioi	soil	
14	*Clitocybe nebularis* (Batsch) P. Kumm.	EP.18-A1580/02-Nov-2018	Vourkoti	leaf litter	

Table 2. *Cont.*

a/a	Species Name	Specimen Code/Collection Date	Locality	Substrate Type	GenBank Accession No.
15	*Clitocybe phyllophila* (Pers.) P. Kumm.	EP.18-A1579/02-Nov-2018	Vourkoti	leaf litter	
16	*Clitopilus hobsonii* (Berk.) P.D.Orton	EP.02-A331/29-Apr-2002	Vori	fallen trunk	
17	*Coniophora puteana* (Schumach.) P. Karst.	EP.11-A1017/5-Jan-2011 EP.17-A1411/24-Nov-2017 EP.19-A1679/02-Dec-2019	Vori Lefka Remata	fallen trunk fallen trunk fallen trunk	MT458507 MT458508 MT458509
18	*Coprinellus disseminates* (Pers.) J.E. Lange	EP.01-A260/26-Dec-2001 EP.19-A1654/29-Nov-2019	Vori Vori	rotten wood around stump	
19	*Coprinellus radians* (Fr.) Vilgalys, Hopple & Jacq. Johnson	EP.19-A1655/29-Nov-2019 EP.19-A1668/02-Dec-2019	Vori Remata	woody residues woody residues	MT458510
20	**Coprinopsis melanthina* (Fr.) Örstadius & E. Larss.	EP.17-A1412/24-Nov-2017 EP.19-A1659/29-Nov-2019	Lefka Vori	woody residues woody residues	MT458511
21	*Coriolopsis gallica* (Fr.) Ryvarden	EP.11-A1016/05-Jan-2011	Vori	standing trunk	
22	*Crepidotus luteolus* (Lambotte) Sacc.	EP.01-A268/26-Dec-2001	Vori	woody residues	
23	*Delicatula integrella* (Pers.) Fayod	EP.19-A1634/16-Nov-2019 EP.19-A1640/17-Nov-2019	Katakalaioi Achlas riv.	trunk base bark living tree	
24	*Entoloma alnicola* Noordel. & Polemis	EP.02-A364/22-Sep-2002	Evrousies	soil	
25	*Entoloma incanum* (Fr.) Hesler	EP.02-A362/22-Sep-2002 EP.04-A674/24-Oct-2004 EP.19-A1633/16-Nov-2019	Evrousies Achlas riv. Katakalaioi	soil soil soil	
26	*Entoloma juncinum* (Kühner & Romagn.) Noordel.	EP.02-A362/22-Sep-2002	Evrousies	soil	
27	*Entoloma mougeotii* (Fr.) Hesler	EP.02-A446/11-Oct-2002	Achlas riv.	soil	
28	**Entoloma uranochroum* Hauskn. & Noordel.	EP.19-A1615/14-Nov-2019	Evrousies	soil/leaf-litter	
29	*Exidiopsis galzinii* (L.S. Olive) K. Wells	EP.02-A448/11-Oct-2002	Achlas riv.	fallen trunk	
30	*Fibroporia citrine* (Bernicchia & Ryvarden) Bernicchia & Ryvarden	EP.17-A1356/11-Nov-2017	Zenio	fallen trunk	
31	*Fomes fomentarius* (L.) Fr.	EP.20-A1681/05-Jan-2020	Vori	standing trunk	
32	*Fuscoporia torulosa* (Pers.) T. Wagner	EP.04-A668/15-Oct-2004 EP.04-A672/24-Oct-2004 EP.17-A1447/26-Nov-2017 EP.19-A1677/02-Dec-2019	Evrousies Achlas riv. Achlas riv. Remata	standing trunk standing trunk standing trunk standing trunk	
33	*Ganoderma adspersum* (Schulzer) Donk	EP.14-A1264/01-Nov-2014 EP.17-A1422/24-Nov-2017	Vori Vori	standing trunk standing trunk	
34	*Ganoderma resinaceum* Boud.	EP.19-A1662/30-Nov-2019	Remata	standing trunk	
35	**Gymnopilus arenophilus* A. Ortega & Esteve-Rav.	EP.03-A659/04-Nov-2003 EP.17-A1408/24-Nov-2017 EP.19-A1674/02-Dec-2019	Apoikia Lefka Remata	rotten stump rotten stump rotten stump	 MT458512 MT458513
36	*Gymnopilus junonius* (Fr.) P.D. Orton	EP.04-A667/15-Oct-2004 EP.04-A825/02-Dec-2004 EP.17-A1385/14-Nov-2017 EP.18-A1587/05-Nov-2018	Apoikia Apoikia Evrousies Vori	standing trunk standing trunk standing trunk standing trunk	
37	*Gymnopus brassicolens* (Romagn.) Antonín & Noordel.	EP.17-A1425/24-Nov-2017 EP.19-A1673/02-Dec-2019	Vori Remata	woody residues woody residues	
38	**Hydropus floccipes* (Fr.) Singer	EP.19-A1658/29-Nov-2019	Vori	standing trunk	
39	*Hyphoderma medioburiense* (Burt) Donk	EP.17-A1361/11-Nov-2017 EP.17-A1381/12-Nov-2017	Zenio Katakalaioi	fallen trunk twigs	 MT458514

Table 2. *Cont.*

a/a	Species Name	Specimen Code/Collection Date	Locality	Substrate Type	GenBank Accession No.
40	**Hyphoderma nemorale* K.H. Larss.	EP.02-A326/29-Apr-2002	Vori	fallen trunk	MT458515
41	*Hyphoderma setigerum* (Fr.) Donk	EP.02-A332/29-Apr-2002 EP.17-A1357/11-Nov-2017	Vori Zenio	fallen trunk fallen trunk	 MT458516
42	**Hyphodermella corrugate* (Fr.) J. Erikss. & Ryvarden	EP.17-A1398/23-Nov-2017	Evrousies	fallen branch	MT458517
43	**Lepiota ochraceofulva* P.D. Orton	EP.19-A1627/15-Nov-2019	Vourkoti	leaf-litter	MT458518
44	* *Lepista ovispora* (J.E. Lange) Gulden	EP.19-A1608/12-Nov-2019	Zenio	leaf-litter	
45	*Lepista nuda* (Bull.) Cooke	EP.18-A1564/01-Nov-2018	Zenio	leaf-litter	
46	*Leucoagaricus melanotrichus* (Malençon & Bertault) Trimbach	EP.19-A1604/12-Nov-2019	Zenio	leaf-litter	
47	*Leucopaxillus gentianeus* (Quél.) Kotl.	EP.18-A1577/01-Nov-2018	Zenio	leaf-litter	
48	*Marasmius rotula* (Scop.) Fr.	EP.02-A360/22-Sep-2002	Evrousies	twigs	
49	*Melanoleuca exscissa* (Fr.) Singer	EP.17-A1415/24-Nov-2017	Vori	soil	MT458519
50	*Merulius tremellosus* Schrad.	EP.04-A680/15-Nov-2004 EP.02-A323/29-Apr-2002	Evrousies Vori	fallen trunk fallen trunk	
51	*Mycena galericulata* (Scop.) Gray	EP.17-A1349/11-Nov-2017 EP.19-A1625/15-Nov-2019 EP.19-A1629/16-Nov-2019 EP.19-A1668/02-Dec-2019	Zenio Vourkoti Katakalaioi Remata	around trunk dead stump around trunk dead stump	 MT458520 MT458521
52	*Mycena haematopus* (Pers.) P. Kumm.	EP.01-A262/26-Dec-2001	Vori	rotten trunk	
53	*Mycena pseudocorticola* Kühner	EP.17-A1449/26-Nov-2017 EP.19-A1667/02-Dec-2019	Achlas riv. Remata	bark living tree bark living tree	
54	*Mycena sanguinolenta* (Alb. & Schwein.) P. Kumm	EP.19-A1642/17-Nov-2019	Achlas riv.	twigs	
55	*Mycetinis scorodonius* (Fr.) A.W. Wilson & Desjardin	EP.19-A1644/17-Nov-2019	Achlas riv.	twigs	
56	*Mycoacia aurea* (Fr.) J. Erikss. & Ryvarden	EP.02-A325/29-Apr-2002 EP.11-A1026/05-Jan-2011	Vori Vori	fallen trunk fallen trunk	
57	*Mycoacia uda* (Fr.) Donk	EP.17-A1351/11-Nov-2017	Zenio	fallen trunk	
58	*Mycoaciella bispora* (Stalpers) Erikss. & Ryvarden	EP.01-A272/26-Dec-2001 EP.02-A330/29-Apr-2002	Vori Vori	fallen trunk fallen trunk	
59	*Paralepista flaccida* (Sowerby) Vizzini	EP.18-A1563/01-Nov-2018 EP.18-A1581/02-Nov-2018	Zenio Vourkoti	leaf-litter leaf-litter	
60	*Parasola kuehneri* (Uljé & Bas) Redhead, Vilgalys & Hopple	EP.18-A1546/22-Feb-2018	Vori	soil	
61	*Peniophora tamaricicola* Boidin & Malenç.	EP.02-A329/29-Apr-2002	Vori	fallen trunk	
62	*Peniophorella praetermissa* (P. Karst.) K.H. Larss.	EP.01-A276/26-Dec-2001 EP.17-A1355/11-Nov-2017	Vori Zenio	fallen trunk fallen trunk	 MT458522
63	*Perenniporia ochroleuca* (Berk.) Ryvarden	EP.02-A461/11-Oct-2002	Achlas riv.	dead wood	
64	**Phanerochaete livescens* (P. Karst.) Volobuev & Spirin	EP.01-A273/26-Dec-2001 EP.18-A1541/22-Feb-2018	Vori Vori	fallen trunk fallen trunk	 MT458523
65	*Phellinus lundellii* Niemelä	EP.17-A1342/13-Apr-2017	Zenio	standing trunk	
66	*Phlebia rufa* (Pers.) M.P. Christ.	EP.17-A1354/11-Nov-2017 EP.20-A1682/05-Jan-2020	Zenio Vori	fallen trunk fallen trunk	MT458524 MT458525

Table 2. *Cont.*

a/a	Species Name	Specimen Code/Collection Date	Locality	Substrate Type	GenBank Accession No.
67	*Phlebiopsis ravenelii* (Cooke) Hjortstam	EP.11-A1018/05-Jan-2011	Vori	fallen trunk	
68	*Phloeomana alba* (Bres.) Redhead	EP.19-A1641/17-Nov-2019	Achlas riv.	bark living tree	MT458526
69	*Phloeomana speirea* (Fr.) Redhead	EP.17-A1378/12-Nov-2017	Lezina	twigs	MT458527
70	*Physisporinus vitreus* (Pers.) P. Karst.	EP.20-A1684/05-Jan-2020	Vori	fallen trunk	
71	*Pilatotrama ljubarskyi* (Pilát) Zmitrovich	EP.18-A1551/22-Feb-2018	Vori	fallen trunk	MT458528
72	*Pleurotus ostreatus* (Jacq.) P. Kumm.	EP.01-A261/26-Dec-2001 EP.18-A1586/05-Nov-2018	Vori Vori	standing trunk standing trunk	
73	*Pluteus cervinus* (Schaeff.) P. Kumm.	EP.01-A321/29-Apr-2002 EP.19-A1623/15-Nov-2019 EP.19-A1676/02-Dec-2019	Vori Vourkoti Remata	dead wood dead wood fallen trunk	MT458529 MT458530
74	*Pluteus nanus* (Pers.) P. Kumm.	EP.19-A1653/29-Nov-2019	Vori	woody residues	
75	*Pluteus salicinus* (Pers.) P. Kumm.	EP.19-A1657/29-Nov-2019	Vori	fallen branch	
76	**Pluteus podospileus* Sacc. & Cub.	EP.19-A1643/17-Nov-2019	Achlas riv.	fallen twigs	
77	*Postia balsamea* (Peck) Jülich	EP.19-A1663/30-Nov-2019	Remata	fallen trunk	
78	*Psathyrella candolleana* (Fr.) Maire	EP.02-A336/04-Jun-2002 EP.19-A1624/15-Nov-2019	Achlas riv. Vourkoti	woody residues woody residues	MT458531
79	*Psathyrella corrugis* (Pers.) Konrad & Maubl.	EP.19-A1601/16-Oct-2019	Vourkoti	soil/buried wood	MT458532
80	**Psathyrella hellebosensis* Deschuyteneer & A. Melzer	EP.17-A1409/24-Nov-2017	Lefka	woody residues	MT458533
81	*Psathyrella microrhiza* (Lasch) Konrad & Maubl.	EP.19-A1603/12-Nov-2019	Vourkoti	woody residues	MT458534
82	*Psathyrella prona* (Fr.) Gill.	EP.02-A363/22-Sep-2002	Evrousies	soil	
83	*Psathyrella vinosofulva* P.D. Orton	EP.17-A1431/24-Nov-2017	Vori	soil	MT458535
84	*Radulomyces confluens* (Fr.) M.P. Christ.	EP.11-A1022/05-Jan-2011	Vori	fallen trunk	
85	*Steccherinum ochraceum* (Pers. ex J.F. Gmel.) Gray	EP.18-A1540/22-Feb-2018	Vori	fallen trunk	
86	*Stereum hirsutum* (Willd.) Pers.	EP.17-A1397/23-Nov-2017 EP.17-A1463/28-Nov-2017 EP.18-A1544/22-Feb-2018 EP.20-A1691/25-Jan-2020	Evrousies Achlas riv. Vori Lefka	trunk/branch trunk/branch trunk branch	
87	*Trechispora nivea* (Pers.) K.H. Larss.	EP.02-A328/29-Apr-2002 EP.20-A1683/05-Jan-2020	Vori Vori	fallen trunk fallen trunk	MT458536
88	*Trametes versicolor* (L.) Lloyd	EP.11-A1019/05-Jan-2011 EP.18-A1542/22-Feb-2018	Vori Vori	fallen trunk standing trunk	
89	*Tubaria furfuracea* (Pers.) Gillet	EP.19-A1652/29-Nov-2019	Vori	woody residues	
90	*Tulostoma fimbriatum* Fr.	EP.17-A1431/24-Nov-2017	Vori	soil	
91	*Vitreoporus dichrous* (Fr.) Zmitr.	EP.02-A445/11-Oct-2002 EP.17-A1439/25-Nov-2017 EP.18-A1538/22-Feb-2017	Achlas riv. Vori Vori	fallen trunk fallen branch fallen branch	
92	*Xylodon raduloides* Riebesehl & Langer	EP.11-A1025/05-Jan-2011 EP.18-A1543/22-Feb-2018 EP.19-A1671/02-Dec-2019	Vori Vori Remata	fallen trunk fallen trunk fallen trunk	MT458538

3.1. The ECM Element

Among the ECM macrofungi recorded (Table 1), the genus *Alnicola* is represented by six species (Figure 3); five of them form part of the sect. *Alnicola* sensu Moreau [45], which is characterized by urticoid cheilocystidia, and one of the sect. *Submelinoideae* Singer with clavate or capitate cheilocystidia [46].

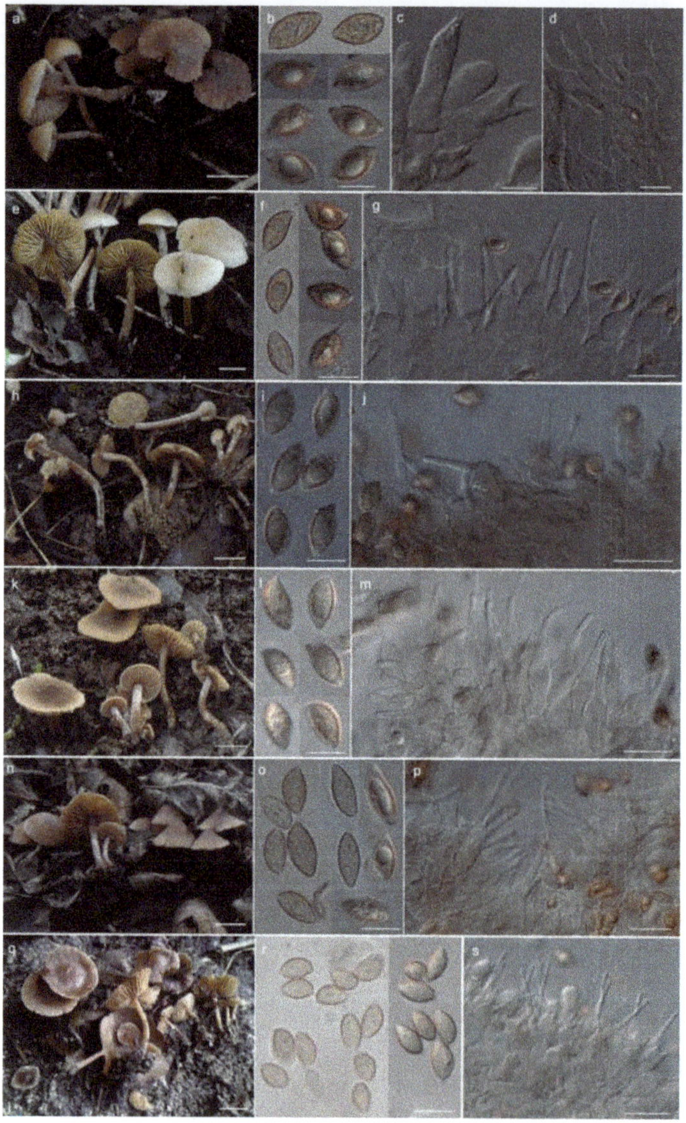

Figure 3. Species of the genus *Alnicola* recorded in Andros alder stands: *A. escharoides* basidiomes, basidiospores and cheilocystidia (**a–c**); *A. umbrina* basidiomes, basidiospores and cheilocystidia (**d–f**); *A. striatula* basidiomes, basidiospores and cheilocystidia (**g–i**); *A. subconspersa* basidiomes, basidiospores and cheilocystidia (**j–l**); *A. luteolofibrillosa* basidiomes, basidiospores and cheilocystidia (**m–o**); *A. inculta* basidiomes, basidiospores, basidia and cheilocystidia (**p–s**). Bars: basidiomes, 1 mm; basidiospores and basidia, 10 μm; cheilocystidia, 20 μm.

Different opinions exist regarding the genus name in pertinent literature since some European authors as well as the Index Fungorum prefer to conserve the name *Naucoria* (Fr.) P. Kumm., whereas Moreau, in his nomenclatural revision, rejected this name in favour of *Alnicola* Kühner [45]; the latter approach is accepted by other European mycologists, the Mycobank, and is also adopted in this work. Moreover, the taxonomy of species of the sect. *Alnicola* remains problematic and, consequently, a phylogenetic analysis was performed to deal with this issue.

The most often found *Alnicola* species in our study were *A. umbrina* (R. Maire) Kühner and *A. escharoides* (Fr.) Romagn., i.e., two of the most common taxa associated with alders in Europe; both constitute new national records for Greece. Particularly *A. escharoides* (syn. *A. citrinella* Moreau & A. de Haan [47]) is distinguished from all other (more or less brownish) species found in Andros by its pale yellowish-buff non striate pileus, the amygdaliform to navicular spores, with prominent ornamentation, measuring 9.9–11.8 × 5.3–5.9 µm, Q = 1.9–2.1 (Figure 3a–c). Following phylogenetic analysis, our specimens are positioned in a distinct group (albeit not adequately supported) together with other sequences from material identified as *A. escharoides* and *A. citrinella* (Figure 4).

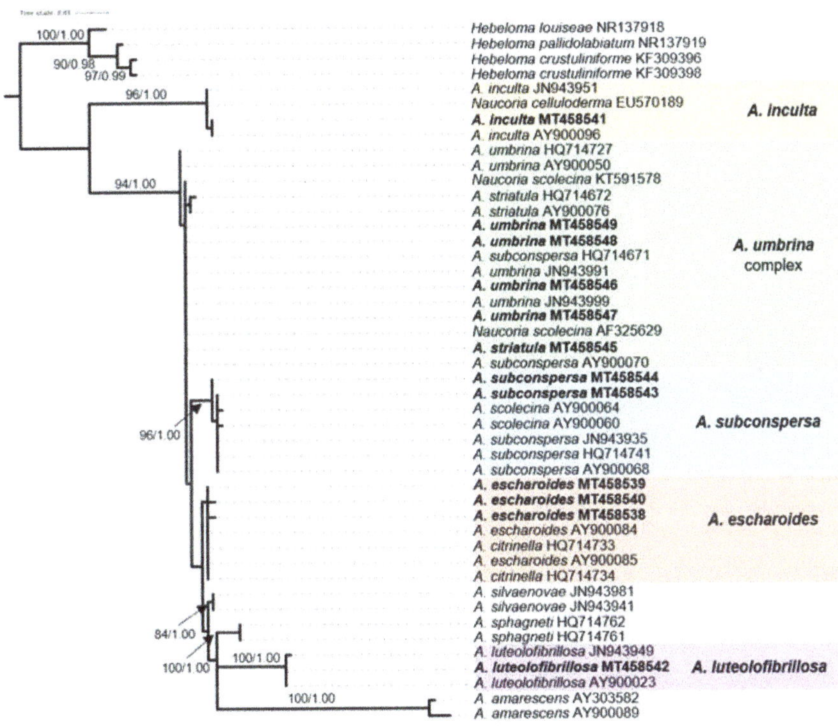

Figure 4. Phylogeny of *Alnicola* species derived from rDNA ITS sequences through ML analysis. Branches are labelled when MLB > 65% and BPP > 0.95. *Hebeloma* species (*H. louiseae*, *H. pallidolabiatum*, *H. crustuliniforme*) were used as outgroups. Boxes include sequences from specimens recorded in the *Alnus glutinosa* habitat.

On the other hand, *A. umbrina* (Figure 3d–f) is hereby considered as a species complex following the nomenclatural concept of Moreau [45] and the outcome of the phylogenetic study by Rochet et al. [19]. According to our observations, *A. umbrina* shows a rather large morphological variability with dark brown hygrophanous pilei bearing prominent striations up to their centre when wet, becoming much lighter and indistinctly striate only at margin when dry. Basidiospores are variable in size and shape, often somewhat elongated fusiform, weakly to moderately verrucose, measuring 10.7–13.6 × 5.2–6.1 μm, Q = 1.9–2.4. Sequences generated in this work clustered together with material identified as *A. umbrina*, *N. scolecina* (Fr.) Quél., *A. striatula* (P.D. Orton) Romagn. and *A. subconspersa* (Kühner ex P.D. Orton) Bon into a group that was not adequately supported (Figure 4). However, the morphological features of specimens identified as *N. scolecina* in Europe are very similar to descriptions of *A. umbrina* [22,33,48,49]. Therefore, *N. scolecina* and *A. umbrina* form part of the same complex and the question whether they constitute different entities or not remains open and in need of further research.

One collection representing another closely related taxon, previously reported as *N. striatula* P.D. Orton (Figure 3g–i) from alder stands in Andros [22], derived from the same site during our recent forays. According to Moreau (2005), *A. striatula* might merely correspond to a pale form of *A. umbrina*, but our morphological studies revealed some noteworthy differences when compared to specimens hereby named *A. umbrina*, i.e., pileus always very prominently striate, smooth and shiny, and (most importantly) significantly smaller basidiospores measuring 8.2–10.0 × 4.5–5.6 μm, Q = 1.7–1.9; these features are in accordance to previous descriptions of *N. striatula* [33,48,50]. As evidenced from our phylogenetic analysis (Figure 4), this particular collection forms part of the *A. umbrina* complex (together with the other two *A. striatula* sequences included in the tree) by using ITS alone; however, since it is morphologically distinct and fits to the widely accepted taxonomic concept of *A. striatula*, we provisionally retain it in this inventory as a separate taxon, until a future multigene approach shows otherwise.

A similar looking species to *A. umbrina*—but less common in Andros—is *A. subconspersa* (Figure 3j–l). The most prominent distinguishing features versus our *A. umbrina* specimens are the non (or very faintly) striate pileus as well as the size and shape of spores, being wider, amygdaliform to navicular and more prominently ornamented, measuring 10.9–12.9 × 6.0–6.8 μm, Q = 1.7–2.0. It is noteworthy that *A. subconspersa* forms a well-supported phylogenetic group including sequences labelled as *A. scolecina* (Fr.) Romagn. (Figure 4), which is indicative of the morphological affinity of these taxa that had apparently led to the development of ambiguous species concepts.

Another collection representing a member of the sect. *Alnicola* was recorded in the alluvial littoral forest of Vori; it corresponds to *A. luteolofibrillosa* Kühner and constitutes the first report of this species in Greece (Figure 3m–o). It is morphologically characterized by non-striate, pale buff, fibrillose to tomentose pilei, with abundant whitish veil remnants on stipe and pileal margin; the respective sequence falls within a highly-supported terminal subgroup corresponding to this species (Figure 4). Lastly, *A. inculta* (Peck) Singer (Figure 3p–s) was recorded only at Zenio (i.e., the site with the highest altitude among those of this study, 850 m) and is reported for the first time in Greece. It forms part of the sect. *Submelinoideae*, and, according to Moreau [45] is conspecific to the taxon widely referred as *N. celluloderma* P.D. Orton, as it is also evidenced by our phylogenetic analysis (Figure 4). Morphologically, this species is easily distinguished from all aforementioned taxa thanks to the clavate to capitate cheilocystidia characterizing members of sect. *Submelinoideae* and the 2-spored basidia.

The most common ECM mushroom in alder stands of Andros belongs to the genus *Paxillus*; it was the first recorded *Alnus*-specific symbiont in the island 25 years ago, and was later repeatedly found in this particular habitat (Figure 5a–d). It was initially identified as *P. rubicundulus* P.D. Orton [22]; however, sequencing of recent collections revealed that it forms part of the newly described taxon *P. olivellus* Moreau P-A, Chaumeton J-P, Gryta H, Jargeat P [51]. Although clearly separated by molecular approaches, *P. olivellus* can be hardly distinguished from *P. rubicundulus* and *P. adelphus* Chaumeton JP, Gryta H, Jargeat P, Moreau P-A on the basis of morphology alone, i.e., only by the olivaceous tinges of the young basidiomes and the basidiospores shape, which are ovoid to ellipsoid in *P. olivellus*, cylindrical in *P. rubicundulus* and short cylindrical in *P. adelphus* [51]. Such features were observed in our specimens since olivaceous tints were always evident in young basidiomes, and spores were ovoid to ellipsoid measuring 6.7–8.1× 4.5–5.2 μm, Q = 1.39–1.66. In addition, ITS sequences from our material originating from various sites in the habitat under study were very similar or identical to those corresponding to *P. olivellus* (including the type), and formed a terminal subgroup with high support (Figure 6). Therefore, this particular species seems to be the only representative of the genus *Paxillus* in the black alder stands of Andros island.

Figure 5. Alder-associated ECM fungi recorded in Andros: *Paxillus olivellus* basidiomes (**a**; bar 1 mm), basidiospores (**b**; bar 10 μm), section of lamella (**c**; bar 20 μm), hymenial cystidium and basidia (**d**; bar 20 μm); *Lactarius obscuratus* basidiomes (**e**, bar 1 mm), basidiospores (**f**, bar 10 μm), pileipellis (**g**, bar 20 μm); *Russula pumila* basidiomes (**h**, bar 1 mm), basidiospores (**i**, bar 10 μm), pileipellis (**j**, bar 20 μm); *Cortinarius americanus* basidiomes (**k**, bar 1 mm); *Inocybe calospora* basidiospores, basidium and pleurocystidium (**l**, bar 10 μm).

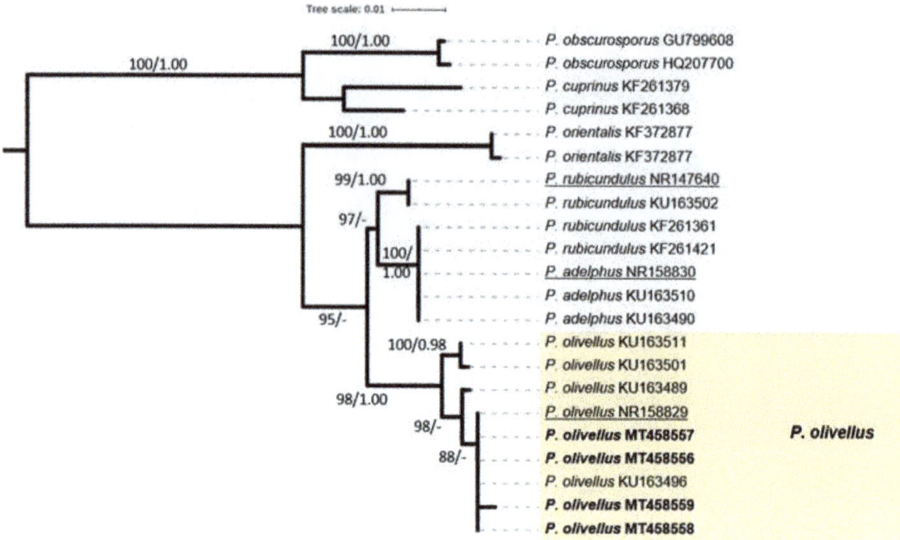

Figure 6. Phylogeny of *Paxillus* species derived from rDNA ITS sequences through ML analysis. Branches are labelled when MLB > 65%, and BPP > 0.95. *P. cuprinus* and *P. obscurosporus* were used as outgroups. The coloured box includes sequences from specimens recorded in the *Alnus glutinosa* habitat.

Lactarius obscuratus (Lasch) Fr. is one the few *Alnus*-specific ECM symbionts of this particular genus; it was found in several inland collection sites dominated by *A. glutinosa*, but not in the alluvial (littoral) forest of Vori (Figure 5e–g). Phylogenetic analysis of our sequences derived from several collections confirmed that they belong to this particular species (Figure 7). However, the respective terminal subgroup in our phylogenetic tree is composed from sequences named either *L. obscuratus* or *L. cyathuliformis* Bon, which is due to the different interpretations existing about this taxon (J. Nuytinck, pers. comm.). In the study of Rochet et al. [19], it is referred as *L. cyathuliformis*, whereas the correct name for the same group is *L. obscuratus* according to Wisitrassameewong et al. [52]. Since the basidiospores average size in our collections (measuring 7.6–8.5 × 6.1–6.3 μm) is in agreement with the concept of *L. obscuratus* (sensu Heilmann-Clausen et al. [31]; according to the same authors, spores of *L. cyathuliformis* have an average size of 8.3–9.9 × 7.0–7.7 μm), we adopt this name for the specimens included in this work. The genus *Russula* Pers. is represented by *R. pumila* Rouzeau & F. Massart (Figure 5h–j) detected in one site only (Vourkoti). Although *R. pumila* is synonymous to *R. alnetorum* Romagn. according to both Index Fungorum and Mycobank, there are different opinions about the synonymy of these two taxa and to the best of our knowledge this issue has not been resolved yet (S. Adamcik, pers. comm.). *R. pumila* is reported to occur mainly in lowlands with *A. glutinosa*, as opposed to *R. alnetorum*, which is mostly recorded in subalpine habitats with *A. viridis* [53,54]; therefore, we adopt the use of the former name.

Tomentella sublilacina (Ellis & Holw.) Wakef. and *T. stuposa* (Link) Stalpers were previously recorded in Andros (in the alluvial alder forest of Vori; [20]) and identified on the basis of their morphology. These names are provisionally retained here due to the absence of precise taxonomic information concerning alder-specific *Tomentella* species. It should be noted that the ITS sequences generated in the frame of this work represent phylogenetically distinct taxa corresponding to entities named "aff. *sublilacina*" and "aff. *stuposa*" in previous studies referring to material originating from alder hosts only [18,55].

It is noteworthy that the first ever report of an alniphilous *Cortinarius* species in Greece derives from a single collection of *C. americanus* A.H. Sm. (Figure 5k), which forms part of a small group of species within the subgenus *Telamonia* (Fr.) Trog known to be associated with *A. glutinosa* and *A. incana* in Europe [33,56]. *C. americanus* is characterized by the minute size, pileus not exceeding 2 cm in diameter, with dark violet to blackish colour, and spores smaller than 10 × 6 μm [33,56]; our collection has spores measuring 7.6–8.4 × 4.9–5.6 μm.

Last, *Inocybe calospora* Quél., recorded for the first time in Greece, was collected in Vourkoti only; it is an easily identified species thanks to its unique star-shaped spiny spores (Figure 5). Among all ECM species included in this inventory it is the only one which is not considered to be exclusively associated with alders, and reported from diverse damp deciduous forests of Europe [33,56].

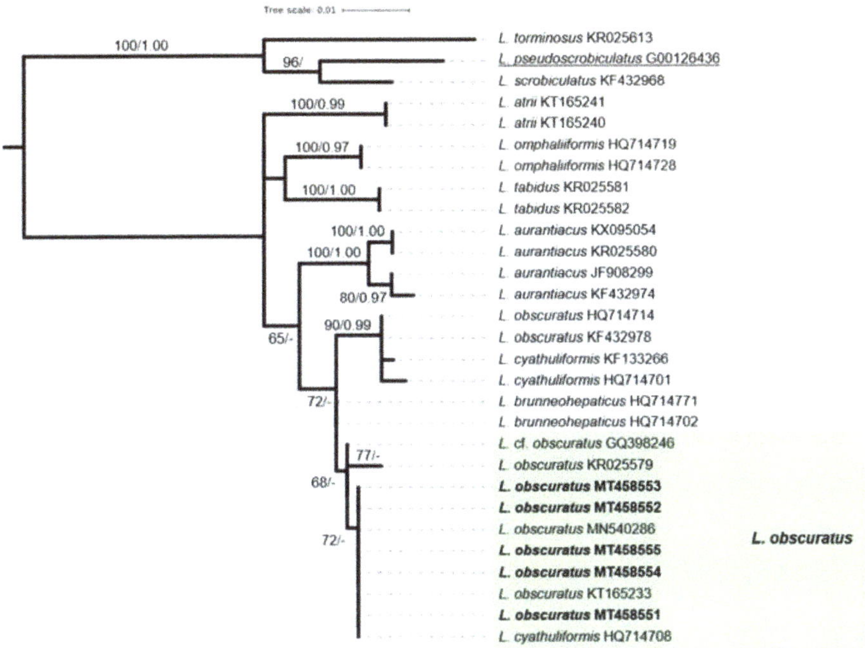

Figure 7. Phylogeny of *Lactarius* species derived from rDNA ITS sequences through ML analysis. Branches are labelled when MLB > 65%, and BPP > 0.95. *L. torminosus*, *L. scrobiculatus* and *L. pseudoscrobiculatus* were used as outgroup. The colored box includes sequences from specimens recorded in the *Alnus glutinosa* habitat.

3.2. The Saproxylic Element

By far the highest number of species recorded in this inventory correspond to white-rot and brown-rot basidiomycetes found on various wood parts of *A. glutinosa* (Table 2). This is quite anticipated since alder trees have a life-span which rarely exceeds 100 years; therefore, they produce large amounts of dead wood. Moreover, in contrast to ECM species, wood-rotting fungi do not show any specificity to alders, with only few exceptions including the common in northern Europe plant-pathogenic polypore *Inonotus radiatus* (Sowerby) P. Karst. [15,57–60], which, however, was not among our findings. Most of the recorded species are wood rotting basidiomycetes that are quite common throughout Europe on deciduous tree species including alders [13,61,62]. The best represented genera of saproxylic fungi were *Hyphoderma* Wallr. (three spp.), *Mycena* (Pers.) Roussel (four spp.), *Pluteus* Fr. (four spp.) and *Psathyrella* (Fr.) Quél. (four spp. on woody residues or buried wood). In total, seven species recorded on dead wood or bark of living alder trees are recorded for the first time in Greece and are presented in more detail below.

Among corticioid basidiomycetes, *Hyphoderma nemorale* K.H. Larss. is a distinct, widely distributed but rare species in Europe [35], which was identified by re-examining an old collection from Vori alluvial forest and further confirmed by ITS sequencing. The presence of thick-walled subicular hyphae and of two types of hymenial cystidia (i.e., short ventricose and subcapitate, and more seldom long tubular with characteristic constrictions, sometimes moniliform, which was the case in our specimen) are the main diagnostic features of this species [63]. *Hyphodermella corrugata* (Fr.) J. Erikss. & Ryvarden (Figure 8a) is fairly common and widespread in Europe [35]. It is easily identified thanks to its characteristic cystidioid hyphal ends appearing in bundles that are heavily incrusted [24]. *Phanerochaete livescens* (P. Karst.) Volobuev & Spirin (Figure 8b) is a species closely related to *Ph. sordida* (P. Karst.) J. Erikss. & Ryvarden which was recently described by using both morphological and phylogenetic criteria [64]. In accordance to the pertinent description, our specimens possessed cystidia with thickened walls to the acute apex, densely covered by crystals, as opposed to the accidentally encrusted, obtuse and thin-walled towards the apex cystidia of *Ph. sordida*. Moreover, the identity of our specimen was confirmed by the respective ITS sequence which was identical to those of *Ph. livescens* as determined by Volobuev et al. [64].

Figure 8. Saproxylic species recorded on Alnus glutinosa wood and litter in Andros: Hyphodermella corrugata (**a**) Phanerochaete livescens (**b**) Pluteus podospileus (**c**) Delicatula integrella (**d**) Coprinopsis melanthina (**e**) Gymnopilus arenophilus (**f**) Hydropus floccipes (**g**) Lepiota ochraceofulva (**h**) Lepista ovispora (**i**) Psathyrella hellebosensis (**j**) Entoloma uranochroum (**k**) Bar: 1 mm.

Among the agaricoid wood-inhabiting fungi, the genus *Pluteus* Fr. is hereby represented by four species, of which *P. podospileus* Sacc. & Cub. (Figure 8c) is reported for the first time in Greece. It belongs to the sect. *Celluloderma* Fay. subsection *Mixtini* Sing. ex Sing, and possesses a pileipellis made up of both fusiform and broadly clavate elements. *P. thomsonii* (Berk. & Broome) Dennis is very similar morphologically but it differs in the absence of pleurocystidia and the shape of cheilocystidia, which are characteristically rostrate [65]. The record of *Delicatula integrella* (Pers.) Rat. (Figure 8d) is worth mentioning since it was reported only once before in Greece, in the content of a regional field-guide [66]. *D. integrella* forms whitish-mycenoid mushrooms of minute size with pileus diameter measuring (in our specimens) 0.4–0.6 cm, reduced, almost vein-like lamellae and non-amyloid, amygdaliform-fusoid spores. It is considered widespread and common in Europe and N. America, and grows on decaying wood and wood debris of deciduous trees [34]. *Coprinopsis melanthina* (Fr.) Örstadius & E. Larss. (Figure 8e) is a striking-looking psathyrelloid species, easily identified due to the relatively large basidiomes with wooly to squamulose pileus (measuring 2–6 cm in diam.) and stipe (up to 6.0 × 0.8 cm), absence of pleurocystidia, almost colourless basidiospores, devoid of germ-pore, measuring 9.8–11.8(13.5) × 5.6–6.5 µm in our collections. This is a rather uncommon European species growing on and around rotten stumps in humid deciduous forests [33]; the only other record of this species in Greece derives from Crete (G. Konstandinidis, pers. comm.).

Gymnopilus arenophilus A. Ortega & Esteve-Rav. (Figure 8f) was described from continental areas of Spain [67] and from maritime dunes in France, under or near Mediterranean pines, on sandy soil by being attached to wood debris or wood, often burnt or buried in the sand [68]. Two of our collections from the alluvial *A. glutinosa* habitat at Lefka were sequenced and found to correspond to this species. Apparently, no native pines exist in Andros while both specimens were growing on rotten alder stumps, a fact that largely expands the so far known ecological and geographical range of this Mediterranean species. Morphologically, our specimens possessed features that fit well to the taxonomic concept of *G. arenophilus* i.e., smooth to fibrillose pileal surface, bitter taste, ellipsoid to subamygdaliform, moderately verrucose spores, measuring 8–10 × 5.5–6.5 µm, lageniform cheilocysidia often with subcapitate apex, 25–45 × 5–8 µm and absence of pleurocystidia. On the other hand, the size of basidiomes was significantly larger, with pilei up to 10 cm in diam. and a sturdy stipe often thicker than 1 cm. One previous collection of ours also found on rotten alder stump, and originally identified as *G. picreus* (Pers.) P. Karst. [22], is now re-assessed as *G. arenophilus*.

Hydropus floccipes (Fr.) Singer (Figure 8g) is a rare mycenoid species generally found to grow on decayed trunks of deciduous trees in damp forests, and is characterized by non-amyloid, subglobose spores, not blackening basidiomes, and typically scabrous stipe with grey-brown spots [33,69]. In addition, our specimens possessed yellowish stipe, previously reported for *H. floccipes* var. *luteipes* Ortega & Zea described from Spain [70]; the latter is otherwise microscopically identical and of unknown phylogenetic status. Two unpublished reports of *H. floccipes* exist from Greece (D. Sofronis and G. Konstandinidis, pers. comm.).

3.3. Litter and Other Terrestrial Decomposers

Apart ECM and saproxylic fungi, several other mushroom species were recorded under black alder trees, and therefore constitute a part of the fungal diversity of the *A. glutinosa* priority habitat in Andros. Needless to say, none of these species is specifically linked to alders; instead they are considered 'generalists' to be found in both deciduous and coniferous forests. Among them, the following four species are recorded for the first time in Greece. *Lepiota ochraceofulva* P.D. Orton (Figure 8h) is a rather rare (but widespread) species in Europe, reported from *Fagus* and other deciduous trees on humus-rich, loamy soil [71,72]; it forms highly toxic mushrooms containing amanitins. Our single collection consisted of few basidiomes growing on a thick layer of leaf-litter under *A. glutinosa.* They are characterized by pilei of up to 7 cm in diam., with orange-brown scales; lamellae forming a distinct collarium and reddish-orange in maturity; spores ellipsoid to oblong, dextrinoid, not metachromatic in cresyl-blue, measuring 5.4–7.5(8.1) × 3.5–4(4.5) µm; basidia (2)4-spored, clamped; cheilocystidia short clavate to cylindrical, rarely papilate, often in chains; pileipellis, a hymeniderm, composed of more or less clavate elements up to 50 μm long.

Lepista ovispora (J.E. Lange) Gulden (Figure 8i) is an uncommon (albeit widespread) European species recorded only once during this study on leaf litter under alders. Typical diagnostic features are the densely caespitose habit, the relatively fleshy basidiomes, the brown hygrophanous pileus with pruinose surface [73]. In addition, our specimens possessed spores ovoid to broadly ellipsoid, finely punctate, 4.7–6.8 × 3.8–4.4 μm, clamped basidia and no cystidia. *Psathyrella hellebosensis* D. Deschuyteneer, A. Melzer (Figure 8j) was recently described from Belgium [74] and was later reported from riparian alder habitats in Italy [75]. The morphological features of our collection are in agreement with the morphology of Belgian and Italian basidiomes, but since it corresponds to a rarely reported species, a detailed description of our material is hereby provided: pileus up to 3 cm in diam., hygrophanous from dark reddish-brown to greyish-beige, with scanty remains of veil; lamellae subdistant with whitish edge; stipe 2–4 × 0.2–0.3 cm, not rooting; spores 7.3–8.7 × 4.5–5.7 μm, Q = 1.43–1.73, ovoid to angular in face-view and not or weakly phaseoliform in side-view, not opaque; lamellae edge sterile composed exclusively of sphaeropendunculate paracystidia (no pleurocystidioid paracystidia were observed); pleurocystidia 34–48 × 11–17 μm, utriform. The material was collected from wet soil by the alluvial stream banks. This species shows high phylogenetic affinity to *P. thujina* A. H. Sm. by using ITS sequences only; however, it is clearly separated when the tef-1α marker is added in the phylogenetic analysis, while it is also distinguished by its distinctly larger and prominently phaseoliform spores [75].

Previous studies on the mycodiversity of Andros island reported the occurrence of four *Entoloma* species, one of them was new to science, i.e., *E. alnicola* Noordel. & Polemis [22,25]. Our recent field work resulted in other interesting collections of *Entoloma* spp. for which the identity and phylogenetic relationships to closely allied taxa are still under investigation. However, by using morphology alone, the presence of a rare European species was confirmed, namely *E. uranochroum* Hauskn. & Noordel. (Figure 8k) recorded for the first time in an alder habitat. This beautiful dark blue-violet mushroom was so far reported from subalpine meadows on calcareous soil in Austria (type locality) and the French Alps. Moreover, its striking microscopical features, e.g., the large fusiform cheilocystidida with granular yellowish-brown content, place it in the distinct section *Ramphocystotae* (Largent) Noordel., together with only one other European representative, namely *E. rhynchocystidiatum* Noordel. & Liiv [76].

4. Conclusions

A long-term study of the diversity of macrofungi in alder stands of Andros resulted in an inventory consisting of 106 species of basidiomycetes, including 21 taxa recorded for the first time in Greece. The majority of findings corresponded to saprotrophs (#92, mainly wood-rotting fungi) and the rest were ECM species. Considering the limited size of the area under study in a small Aegean island, the outcome of this work in terms of the number of taxa and variability is indicative of the wealth of the *A. glutinosa* priority habitat. However, the black alder stands in Andros have suffered considerably from floods in the past (as a consequence of fires that destroyed vegetation in the surrounding mountains which acted as a physical barrier protecting from downhill water runoffs) and their regeneration is hindered due to grazing by feral goats. The importance of fungi in the conservation/restoration of such natural habitats was demonstrated in the past [77,78], and recent activities focus at improving the status of the degenerated alder stands by exploiting indigenous ECM fungi as inoculants to young alder seedlings prior to their transplantation on site. Moreover, new knowledge about mushroom diversity and the ecological role of this group of organisms seems to enhance considerably people's perception and awareness, and hence facilitates implementation of conservations actions which are currently under way in selected alder stands of Andros.

Supplementary Materials: The following is available online at http://www.mdpi.com/1424-2818/12/6/232/s1, Table S1: Details of the 10 sampling sites in Andros island from where basidiomes were collected: locality name, coordinates, altitude (m a.s.l.) and surface of the study area (m^2).

Author Contributions: Conceptualization, E.P. and G.Z.; methodology, E.P., G.I.Z., V.D. and V.F.; validation, E.P., V.D. and V.F.; formal analysis, E.P., G.I.Z. and V.F.; investigation, E.P., G.I.Z., V.D. and V.F.; data curation, E.P., G.I.Z. and V.F.; writing—original draft preparation, E.P.; writing—final draft, G.I.Z.; review and editing—final draft, E.P., G.I.Z., V.D. and V.F.; supervision, G.I.Z.; project administration, G.I.Z.; and funding acquisition, G.I.Z. All authors have read and agreed to the published version of the manuscript.

Funding: This study was funded by the project titled "Conservation of priority species and habitats of Andros Island protected area integrating socioeconomic considerations" (European Commission – LIFE-Nature, LIFE16 NAT/GR/000606).

Acknowledgments: We would like to thank V. Goritsas for the preparation of the map figures included in this work, and S. Adamcik, B. Dima, M. Noordeloos and J. Nuytinck for helpful discussions on some of the findings of this study.

Conflicts of Interest: The authors declare no conflict of interest.

References

1. Kajba, D.; Gračan, J. *EuFORGEN Technical Guidelines for Genetic Conservation and Use for Black Alder (Alnus glutinosa)*; Bioversity International: Rome, Italy, 2003; pp. 1–6.
2. Karl, T.R.; Trenberth, K.E. Modern Global Climate Change. *Science* **2003**, *302*, 1719–1723. [CrossRef] [PubMed]
3. McEwan, N.R.; Wilkinson, T.; Girdwood, S.E.; Snelling, T.J.; Collins, T.; Dougal, K.; Jones, D.L.; Godbold, D.L. Evaluation of the microbiome of decaying alder nodules by next generation sequencing. *Endocyt. Cell Res.* **2017**, *28*, 14–19.
4. Roy, M.; Pozzi, A.C.; Gareil, R.; Nagati, M.; Manzi, S.; Nouioui, I.; Sharikadze, N.; Jargeat, P.; Gryta, H.; Moreau, P.-A.; et al. Alder and the Golden Fleece: High diversity of *Frankia* and ectomycorrhizal fungi revealed from *Alnus glutinosa* subsp. *barbata* roots close to a Tertiary and glacial refugium. *PeerJ* **2017**, *5*, e3479. [CrossRef] [PubMed]
5. Orfanoudakis, M.Z.; Hooker, J.E.; Wheeler-Jones, C.T. Early interactions between arbuscular mycorrhizal fungi and *Frankia* during colonisation and root nodulation of *Alnus glutinosa*. *Symbiosis* **2004**, *36*, 69–82.
6. Põlme, S.; Öpik, M.; Moora, M.; Zobel, M.; Kohout, P.; Oja, J.; Kõljalg, U.; Tedersoo, L. Arbuscular mycorrhizal fungi associating with roots of *Alnus* and *Rubus* in Europe and the Middle East. *Fungal Ecol.* **2016**, *24*, 27–34. [CrossRef]
7. Harley, J.L.; Smith, S.E. *Mycorrhizal Symbiosis*; Academic Press: London, UK, 1983; pp. 1–483.
8. Pritsch, K.; Munch, J.C.; Buscot, F. Characterization and identification of black alder ectomycorrhizas by PCR/RFLP analyses of the rDNA internal described spacer (ITS). *New Phytol.* **1997**, *137*, 357–369. [CrossRef]
9. Pritsch, K.; Munch, J.C.; Buscot, F. Morphological and anatomical characterisation of black alder *Alnus glutinosa* (L.) Gaertn. ectomycorrhizas. *Mycorrhiza* **1997**, *7*, 201–216. [CrossRef]
10. Boyle, H. Aspekte der Macromycetenflora dreier Erlenbrücher Norddeutschlands und vergleichende PCR/RFLP- Analyse ausgewählter ectomycorrhizaler Mycobionten. *EcoSys Suppl.* **1996**, *10*, 1–106.
11. Brunner, I.; Horak, E. Mycoecological analysis of *Alnus* associated macrofungi in the region of the Swiss National Park as recorded by J. Favre (1960). *Mycol. Helv.* **1990**, *4*, 111–139.
12. Bujakiewicz, A.M. Macrofungi in the alder and alluvial forests in various parts of Europe and North America. *Opera Bot.* **1989**, *100*, 29–41.
13. Kunttu, P.; Kotiranta, H.; Kulju, M.; Pasanen, H.; Kouki, J. Occurrence patterns, diversity and ecology of aphyllophoroid fungi on the black alder (*Alnus glutinosa*) in an archipelago in the Baltic Sea. *Ann. Bot. Fenn.* **2016**, *53*, 173–193. [CrossRef]
14. Senn-Irlet, B.; Mürner, R.; Martini, E.; Küffer, N.; de Marchi, R.; Bieri, G. Saprobic fungi on wood and litter of *Alnus alnobetula* in the Swiss Alps. *Mycotaxon* **2012**, *120*, 506.
15. Strid, Å. Wood-inhabiting fungi of alder forests in north-central Scandinavia 1. Aphylloporales (Basidiomycetes). Taxonomy, ecology and distribution. *Wahlenbergia* **1975**, *1*, 1–237.
16. Griesser, B. Mykosoziologie der Grauerlen-und Sanddorn-Auen (Alnetum incanae, Hippophaëtum) am Hinterrhein (Domleschg, Graubünden, Schweiz). *Ver. Geobot. Inst. ETH* **1992**, *109*, 1–235.
17. Molina, R. Ectomycorrhizal specificity in the genus *Alnus*. *Can. J. Bot.* **1981**, *59*, 325–334. [CrossRef]

18. Tedersoo, L.; Suvi, T.; Jairus, T.; Ostonen, I.; Põlme, S. Revisiting ectomycorrhizal fungi of the genus *Alnus*: Differential host specificity, diversity and determinants of the fungal community. *New Phytol.* **2009**, *182*, 727–735. [CrossRef]
19. Rochet, J.; Moreau, P.-A.; Manzi, S.; Gardes, M. Comparative phylogenies and host specialization in the alder ectomycorrhizal fungi *Alnicola*, *Alpova* and *Lactarius* (Basidiomycota) in Europe. *BMC Evol. Biol.* **2011**, *11*, 40. [CrossRef]
20. Dimou, D.M.; Polemis, E.; Zervakis, G.I. Macromycetes associated with *Alnus glutinosa* in Greece. *Phytopathol. Mediterr.* **2006**, *45*, 78.
21. Dimou, D.M.; Zervakis, G.I.; Polemis, E. Mycodiversity studies in selected ecosystems of Greece: IV. Macrofungi from *Abies cephalonica* forests and other intermixed tree species. (Oxya mountain, central Greece). *Mycotaxon* **2008**, *104*, 39–42.
22. Polemis, E.; Dimou, D.M.; Tzanoudakis, D.; Zervakis, G.I. Diversity of Basidiomycota (subclass Agaricomycetidae) in the island of Andros (Cyclades, Greece). *Nova Hedwig.* **2012**, *95*, 25–58. [CrossRef]
23. Polemis, E.; Dimou, D.; Zervakis, G.I. The family *Hymenochaetaceae* (Agaricomycetes, Basidiomycota) in the islands of the Aegean Archipelago (Greece). *Plant Biosyst.* **2013**, *147*, 306–314. [CrossRef]
24. Polemis, E.; Roberts, P.; Dimou, D.M.; Zervakis, G.I. Heterobasidiomycetous fungi from Aegean Islands (Greece): New annotated records for a neglected group. *Plant Biosyst.* **2016**, *150*, 295–303. [CrossRef]
25. Noordeloos, M.; Polemis, E. Studies in the genus *Entoloma* (Basidiomycetes, Agaricales) from the Kiklades (C. Aegean, Greece). *Mycotaxon* **2008**, *105*, 301–312.
26. Eriksson, J.; Ryvarden, L. *The Corticiaceae of North Europe, Vol. 4: Hyphodermella—Mycoacia*; Fungiflora: Oslo, Norway, 1976; pp. 549–886.
27. Bas, C.; Kuyper, T.W.; Noordeloos, M.E.; Vellinga, E.C. (Eds.) *Flora Agaricina Neerlandica*; A. A. Balkema: Rotterdam, The Netherlands, 1990; Volume 2, pp. 1–137.
28. Bas, C.; Kuyper, T.W.; Noordeloos, M.E.; Vellinga, E.C. (Eds.) *Flora Agaricina Neerlandica*; A. A. Balkema: Rotterdam, The Netherlands, 1995; Volume 3, pp. 1–183.
29. Bas, C.; Kuyper, T.W.; Noordeloos, M.E.; Vellinga, E.C. (Eds.) *Flora Agaricina Neerlandica*; A. A. Balkema: Rotterdam, The Netherlands, 1999; Volume 4, pp. 1–191.
30. Noordeloos, M.E.; Kuyper, T.W.; Vellinga, E.C. (Eds.) *Flora Agaricina Neerlandica*; A. A. Balkema: Rotterdam, The Netherlands, 2001; Volume 5, pp. 1–170.
31. Heilmann-Clausen, J.; Verbeken, A.; Vesterholt, J. *The Genus Lactarius. Fungi of Northern Europe*; Danish Mycological Society, Svampetryk: Copenhagen, Denmark, 1998; Volume 2, pp. 1–287.
32. Bernicchia, A. *Polyporaceae S.L.*; Candusso: Alassio, Italy, 2005; pp. 1–808.
33. Knudsen, H.; Vesterholt, J. (Eds.) *Funga Nordica. Agaricoid, Boletoid and Cypheloid Genera*; Nordsvamp: Copenhagen, Denmark, 2008; pp. 1–965.
34. Antonín, V.; Noordeloos, M.E. *A Monograph of Hemimycena, Delicatula, Fayodia, Gamundia, Myxomphalia, Resinomycena, Richenella and Pseudomphalina (Tribus Mycenae Sensu Singer, Mycena Excluded)*; IHW Verlag: Eching, Germany, 2004; pp. 1–279.
35. Bernicchia, A.; Gorjón, S.P. *Corticiaceae S.L.*; Candusso: Alassio, Italy, 2010; pp. 1–1008.
36. Aronsen, A.; Læssøe, T. *The Genus Mycena. Fungi of Northern Europe*; Svampetryk: Copenhagen, Denmark, 2016; Volume 5, pp. 1–373.
37. White, T.J.; Bruns, T.; Lee, S.; Taylor, J.W. Amplification and direct sequencing of fungal ribosomal RNA genes for phylogenetics. In *PCR Protocols: A Guide to Methods and Applications*; Innis, M.A., Gelfand, D.H., Sninsky, J.J., White, T.J., Eds.; Academic Press Inc.: New York, NY, USA, 1990; pp. 315–322.
38. Hall, T.A. BioEdit: A User-Friendly Biological Sequence Alignment Editor and Analysis Program for Windows 95/98/NT. *Nucleic Acids Symp. Ser.* **1999**, *41*, 95–98.
39. Katoh, K.; Rozewicki, J.; Yamada, K.D. MAFFT online service: Multiple sequence alignment, interactive sequence choice and visualization. *Brief Bioinform.* **2019**, *20*, 1160–1166. [CrossRef]
40. Kumar, S.; Stecher, G.; Li, M.; Knyaz, C.; Tamura, K. MEGA X: Molecular Evolutionary Genetics Analysis across computing platforms. *Mol. Biol. Evol.* **2018**, *35*, 1547–1549. [CrossRef] [PubMed]
41. Stamatakis, A.; Hoover, P.; Rougemont, J. A rapid bootstrap algorithm for the RAxML Web servers. *Syst. Biol.* **2008**, *57*, 758–771. [CrossRef]

42. Ronquist, F.; Teslenko, M.; van der Mark, P.; Ayres, D.L.; Darling, A.; Höhna, S.; Larget, B.; Liu, L.; Suchard, M.A.; Huelsenbeck, J.P. MrBayes 3.2: Efficient Bayesian phylogenetic inference and model choice across a large model space. *Syst. Biol.* **2012**, *61*, 539–542. [CrossRef]
43. Darriba, D.; Taboada, G.L.; Doallo, R.; Posada, D. jModelTest 2: More models, new heuristics and parallel computing. *Nat. Methods* **2012**, *9*, 772. [CrossRef] [PubMed]
44. Letunic, I.; Bork, P. Interactive tree of life (iTOL) v3: An online tool for the display and annotation of phylogenetic and other trees. *Nucleic Acids Res.* **2016**, *44*, W242–W245. [CrossRef] [PubMed]
45. Moreau, P.-A. A nomenclatural revision of the genus *Alnicola* (Cortinariaceae). *Fungal Divers.* **2005**, *20*, 121–155.
46. Moreau, P.-A.; Peintner, U.; Gardes, M. Phylogeny of the ectomycorrhizal mushroom genus *Alnicola* (Basidiomycota, Cortinariaceae) based on rDNA sequences with special emphasis on host specificity and morphological characters. *Mol. Phylogen. Evol.* **2006**, *38*, 794–807. [CrossRef]
47. De Haan, A.; Moreau, P.-A. Waarnemingen in het genus *Alnicola* (Zompzwam) in Vlaanderen (3). *Steerbeckia* **2012**, *31*, 3–15.
48. Moser, M. *Keys to Agarics and Boleti*; R. Phillips: Tonbridge, UK, 1983; pp. 1–535.
49. Horak, E. *Röhrlinge und Blätterpilze in Europa—Unter der Mitarbeit von Anton Hausknecht (Bolbitiaceae) und P.A. Moreau (Alnicola)*; Elsevier Spektrum Akademischer: Heidelberg, Germany, 2005; pp. 1–557.
50. Henrici, A. Keys to *Naucoria* in Britain. *Field Mycol.* **2009**, *9*, 55–62. [CrossRef]
51. Jargeat, P.; Moreau, P.-A.; Gryta, H.; Chaumeton, J.P.; Gardes, M. *Paxillus rubicundulus* (Boletales, Paxillaceae) and two new alder-specific ectomycorrhizal species, *Paxillus olivellus* and *Paxillus adelphus*, from Europe and North Africa. *Fungal Biol.* **2016**, *120*, 711–728. [CrossRef]
52. Wisitrassameewong, K.; Looney, B.P.; Le, H.T.; De Crop, E.; Das, K.; Van de Putte, K.; Eberhardt, U.; Jiayu, G.; Stubbe, D.; Hyde, K.D.; et al. *Lactarius* subgenus *Russularia* (Basidiomycota, Russulales): Novel Asian species, worldwide phylogeny and evolutionary relationships. *Fungal Biol.* **2016**, *120*, 1554–1581. [CrossRef] [PubMed]
53. Galli, R. *Le Russule. Atlante Pratico—Monografico per la Determinazione delle Russule*; Dalla Natura: Milano, Italy, 2003; pp. 1–480.
54. Floriani, M.; Partacini, G. Sull'identità di *Russula puellaris* var. *leprosa*. *Boll. Gruppo Micol. G. Bres.* **1998**, *40*, 213–218.
55. Roy, M.; Rochet, J.; Manzi, S.; Jargeat, P.; Gryta, H.; Moreau, P.-A.; Gardes, M. What determines *Alnus*-associated ectomycorrhizal community diversity and specificity? A comparison of host and habitat effects at a regional scale. *New Phytol.* **2013**, *198*, 1228–1238. [CrossRef] [PubMed]
56. Breitenbach, J.; Kranzlin, F. *Fungi of Switzerland*; Verlag Mykologia: Lucerne, Switzerland, 2000; Volumes 1–5, pp. 1–342.
57. Yurchenko, E.O. Natural substrata for corticioid fungi. *Acta Mycol.* **2006**, *41*, 113–124. [CrossRef]
58. Ainsworth, M. Some British alder-associated wood- inhabiting fungi. *Field Mycol.* **2010**, *11*, 10–15. [CrossRef]
59. Kotiranta, H.; Saarenoksa, R.; Kytövuori, I. Aphyllophoroid fungi of Finland. A check-list with ecology, distribution, and threat categories. *Norrlinia* **2009**, *19*, 1–223.
60. Niemelä, T.; Kotiranta, H. Polypore survey of Finland 3. The genera *Coltricia*, *Inonotopsis*, *Inonotus* and *Onnia*. *Karstenia* **1983**, *23*, 15–25. [CrossRef]
61. Küffer, N.; Senn-Irlet, B. Diversity and ecology of corticioid basidiomycetes in green alder stands in Switzerland. *Nova Hedwig.* **2000**, *71*, 131–143.
62. Piętka, J.; Grzywacz, A. Grzyby wielkoowocnikowe stwierdzone na olszy czarnej *Alnus glutinosa* (L.) Gaertn. w drzewostanach olszowych wykazujących objawy zamierania. *Sylwan* **2018**, *162*, 22–31.
63. Larsson, K.-H. Two new species in *Hyphoderma*. *Nord. J. Bot.* **1998**, *18*, 121–127. [CrossRef]
64. Volobuev, S.; Okun, M.; Ordynets, A.; Spirin, V. The *Phanerochaete sordida* group (Polyporales, Basidiomycota) in temperate Eurasia, with a note on *Phanerochaete pallida*. *Mycol. Prog.* **2015**, *14*, 1–13. [CrossRef]
65. Vellinga, E.C. Pluteaceae Kolt. & P. In *Flora Agaricina Neerlandica*; Bas, C., Noordeloos, M.E., Kuyper, T.W., Vellinga, E.C., Eds.; A. A. Balkema: Rotterdam, The Netherlands, 1990; Volume 2, pp. 31–64.
66. Konstantinidis, G. *1000 Mushrooms of Western Makedonia*; The Mushroom Friends Society of Western Makedonia: Kastoria, Greece, 2006; pp. 1–523. (In Greek)
67. Ortega, A.; Esteve-Raventós, F. A new species of *Gymnopilus* (Cortinariaceae) from sandy soils in *Pinus* forests. *Persoonia* **2005**, *18*, 505–510.

68. Guzman-Davalos, L.; Ortega, A.; Contu, M.; Vizzini, A.; Rodriguez, A.; Villalobos-Arambula, A.R.; Santerre, A. *Gymnopilus maritimus* (Basidiomycota, Agaricales), a new species from coastal psammophilous plant communities of northern Sardinia, Italy, and notes on *G. arenophilus*. *Mycol. Prog.* **2009**, *8*, 195–205. [CrossRef]
69. Bas, C. *Hydropus* (Kühner) ex Singer. In *Flora Agaricina Neerlandica*; Bas, C., Kuyper, T.W., Noordeloos, M.E., Vellinga, E.C., Eds.; A. A. Balkema: Rotterdam, The Netherlands, 1999; Volume 4, pp. 166–172.
70. Ortega, A.; Zea, M. *Hydropus floccipes* var. *luteipes* Ortega & Zea var. nov. en España meridional. *Bol. Soc. Micol. Madr.* **1991**, *15*, 189–191.
71. Vellinga, E.C. *Lepiota* (Pers.: Fr.) S.F. Gray. In *Flora Agaricina Neerlandica*; Noordeloos, M.E., Kuyper, T.W., Vellinga, E.C., Eds.; A. A. Balkema: Rotterdam, The Netherlands, 2001; Volume 5, pp. 109–151.
72. Gierczyk, B.; Kujawa, A.; Szczepkowski, A.; Chachuła, P. Rare species of *Lepiota* and related genera. *Acta Mycol.* **2011**, *46*, 137–178. [CrossRef]
73. Noordeloos, M.E.; Bas, C. *Flora Agaricina Neerlandica*; Bas, C., Kuyper, T.W., Noordeloos, M.E., Vellinga, E.C., Eds.; A. A. Balkema: Rotterdam, The Netherlands, 1995; Volume 3, pp. 67–75.
74. Deschuyteneer, D.; Melzer, A. *Psathyrella hellebosensis*, a new species from Belgium. *Bull. AMFB* **2017**, *10*, 3–10.
75. Voto, P.; Dovana, F.; Garbelotto, M. A revision of the genus *Psathyrella*, with a focus on subsection *Spadiceogriseae*. *FUSE* **2019**, *4*, 97–170. [CrossRef] [PubMed]
76. Noordeloos, M.E. *Entoloma s.l. Supplemento. Fungi Europei Vol. 5a*; Candusso: Alassio, Italy, 2004; pp. 1–760.
77. Glassman, S.; Levine, C.; DiRocco, A.; Battles, J.J.; Bruns, T.D. Ectomycorrhizal fungal spore bank recovery after a severe forest fire: Some like it hot. *ISME J.* **2016**, *10*, 1228–1239. [CrossRef] [PubMed]
78. Zotti, M.; Persiani, A.M.; Ambrosio, E.; Vizzini, A.; Venturella, G.; Donnini, D.; Angelini, P.; Di Piazza, S.; Pavarino, M.; Lunghini, D.; et al. Macrofungi as ecosystem resources: Conservation versus exploitation. *Plant Biosyst.* **2013**, *147*, 219–225. [CrossRef]

diversity

MDPI

Article

Genetic Structure and Phylogeography of *Tuber magnatum* Populations

Beatrice Belfiori [1], Valentina D'Angelo [2], Claudia Riccioni [1], Marco Leonardi [2], Francesco Paolocci [1], Giovanni Pacioni [2] and Andrea Rubini [1,*]

1 National Research Council, Institute of Biosciences and Bioresources–Perugia Division, Via Madonna Alta n. 130, 06128 Perugia, Italy; beatrice.belfiori@ibbr.cnr.it (B.B); claudia.riccioni@ibbr.cnr.it (C.R.); francesco.paolocci@ibbr.cnr.it (F.P.)

2 Department of Life, Health & Environmental Sciences, University of L'Aquila, 67010 Coppito-L'Aquila, Italy; valentina.dangelo@accordphoenix.it (V.D.); marco.leonardi@univaq.it (M.L.); giovanni.pacioni@univaq.it (G.P.)

* Correspondence: andrea.rubini@ibbr.cnr.it

Received: 31 December 2019; Accepted: 22 January 2020; Published: 24 January 2020

Abstract: The ectomycorrhizal fungus *Tuber magnatum* produces the white truffle appreciated worldwide for its unique aroma. With respect to other *Tuber* spp. of economic interest, *T. magnatum* presents a narrower geographical range. This species has, in fact, long been considered endemic to Italy. However, over the last few decades several reports have documented the presence of white truffles in different Mediterranean countries and in particular in various areas of south-east Europe. In this study, samples from several Pannonian and Balkan countries such as Hungary, Serbia, Romania, Bulgaria and Greece have been collected and genotyped with microsatellite markers and the data merged with those available for Italian populations. Our objectives were to test whether Italian and south-east European populations are differentiated and to evaluate the genetic diversity of *T. magnatum* all over its distributional range. We show the genetic structure of *T. magnatum* populations with the differentiation of four main groups: northern Italy, central-northern Italy, southern Italy and the Balkan/Pannonian region. The present study allowed us to refine the evolutionary history of *T. magnatum* and track the possible post-glacial expansion route of this species. The assessment of *T. magnatum*'s genetic structure is not only of scientific relevance, but it is also important for the conservation and market traceability of this prestigious fungus.

Keywords: microsatellite; SSR; white truffle; genetic diversity

---

1. Introduction

Species of the genus *Tuber* (Ascomycota, Pezizales, Tuberaceae) establish symbiosis with the roots of several tree and shrub species by forming structures for nutrients exchange, known as ectomycorrhiza [1]. In virtue of this mutualistic relationship these fungi produce hypogeous fruiting bodies (ascomata) known as truffles, that produce their spores sequestered within the surrounding tissues [2]. *Tuber* spp. thus rely on mycophagists for spore dispersal. Truffles produced by several *Tuber* spp. hold distinctive aromatic properties, which make them appreciated and marketed worldwide as food delicacies. Among edible *Tuber* spp. the black truffles harvested in Europe (*T. melanosporum* Vittad. and *T. brumale* Vittad.), the black summer truffle *T. aestivum* Vittad., the whitish truffle *T. borchii* Vittad. and the white truffle *T. magnatum* Pico are of particular relevance.

The evaluation of the intraspecific genetic diversity and population genetic structure of a species is crucial to understand its biology and ascertain its origin, history and evolution. By using molecular markers and performing a wide geographical sampling, a fine assessment of the population genetic structure of *T. melanosporum*, *T. brumale*, *T. indicum*, *T. aestivum* and *T. magnatum* has been performed [3–9].

These studies highlighted the presence of geographically structured populations and phylogeographic patterns in these species.

Regarding *T. magnatum*, its genetic variability was initially investigated over a low number of specimens collected in Italy [10,11] or Italy and Croatia [12], using RAPD (Random Amplification of Polymorphic DNA), ITS-RFLP (Restriction Fragment Length Polymorphism of the Internal Transcribed Spacer of rDNA) and SNPs (Single Nucleotide Polymorphism) either in the ITS region, in the ß-tubulin gene or in a SCAR (Sequence Characterized Amplified Region). All these studies showed a very limited intraspecific polymorphism [13]. A broader investigation was carried out by Frizzi et al. [14] who analyzed the polymorphism of eleven isoenzymes on 139 specimens from 13 Italian populations. The low genetic variability across populations and the lack of an interpretable evolutionary trajectory were thought to be in agreement with a self-reproductive system, a restricted species endemism and a relatively recent differentiation of this taxon [14]. A few years later, by employing seven polymorphic simple sequence repeats (SSR) loci over 316 specimens from Italy and the Istrian peninsula (Croatia and Slovenia), Rubini and colleagues [4] disclosed for the first time an isolation by distance pattern and a phylogeographic structure in *T. magnatum*, with central Italy that likely represented a refugium for this species during the last ice age. In addition, this study has been instrumental for a deep reevaluation of the life cycle and the reproductive strategies of all *Tuber* spp. and to prove that truffle ascocarps are mainly made of the haploid, maternal tissue [15].

Around the late 90s of the last century, it became clear that *T. magnatum* can also be found in the Balkan peninsula and countries nearby [16] and, although more sporadically, in the south-east of France [17] and Switzerland [18]. The recent discovery of natural *T. magnatum* populations in areas ranging from Greece until Hungary and Romania ([13], and references therein), calls now for studies based on a larger sampling area than before. On these premises, here we employed SSR markers and an extensive sampling on most of the *T. magnatum* distributional range to shed more light on the genetic structure and phylogeography of this species. In particular, we aimed at evaluating whether Italian and south-east European populations are genetically differentiated and tracking the post-glacial expansion pattern of this species. The Balkan peninsula, like the Italian one, in fact could have represented a glacial refugia for *T. magnatum* during the last glaciation.

The assessment of the genetic diversity distribution could reveal important findings for aspects spanning from ecology, conservation and marketing of this prestigious fungus.

2. Materials and Methods

2.1. Sample Source and DNA Analysis

Tuber magnatum samples were collected in 2015–2016 with the help of local pickers. Sampling locations were mainly in Pannonian and Balkan countries and more marginally in central and southern Italy (Table 1). Genomic DNA was isolated from freeze-dried and fresh ascocarps (1–5 mg) according to Paolocci et al. [19]. DNA quantity and quality were evaluated using the spectrophotometer Nanodrop (MySpec, Wilmington, DE, USA). The isolated DNA was diluted to 20 ng/μL and stored at −20 °C. All samples were genotyped using eight microsatellite loci previously characterized in this species: the loci MA4, MA7, MA14, MA12, MA19 and MA13 derived from Rubini et al. [20] and the loci MA2-1 and MA5-1 derived from Rubini et al. [4]. The SSR loci were PCR-amplified using multiplex PCR [19]. To this purpose, two panels, each consisting of four loci, were defined (Table S1). PCR conditions were those reported in Rubini et al. [20]. Moreover, the locus MA13 was analyzed in all samples considered by Rubini et al. [4].

The SSR amplicons were analyzed by capillary electrophoresis using an ABI 3130 Genetic Analyzer in presence of the Genescan 500 LIZ size standard (Applied Biosystems, Foster City, CA, USA). Sizing of amplicons and allele scoring were performed using GeneMapper software version 3.7 (Applied Biosystems, Foster City, CA, USA).

Table 1. *Tuber magnatum* samples considered in this study.

Population No.	Population Name (Sampling Location)	Locality	Sample Size*	Mean No. of Alleles**	Expected Heterozygosity**	Allelic Richness**
1	Valle del Crati	Italy (Calabria)	3	1.5 ± 0.19	0.22 ± 0.08	ND^{++}
2	Potenza	Italy (Basilicata)	14 (12)	2.9 ± 0.35	0.37 ± 0.06	1.81 ± 0.15
3	Benevento, Avellino	Italy (Campania)	(5)	2.3 ± 0.37	0.39 ± 0.09	2.05 ± 0.29
4	Isernia, Campobasso	Italy (Molise)	(16)	3.4 ± 0.53	0.56 ± 0.06	2.27 ± 0.18
5	Agnone	Italy (Molise)	(10)	3.4 ± 0.73	0.46 ± 0.11	2.21 ± 0.33
6	Castel del Giudice	Italy (Molise)	(16)	2.9 ± 0.48	0.42 ± 0.09	1.95 ± 0.23
7	Ateleta, SP. Avellana	Italy (Abruzzo-Molise)	36 (18)	4.6 ± 0.89	0.50 ± 0.10	2.17 ± 0.26
8	Roio del Sangro	Italy (Abruzzo-Molise)	(10)	2.9 ± 0.48	0.41 ± 0.09	1.98 ± 0.23
9	Quadri, Rosello	Italy (Abruzzo)	(14)	2.9 ± 0.58	0.44 ± 0.11	2.04 ± 0.27
10	Valle Roveto	Italy (Abruzzo)	(7)	2.3 ± 0.41	0.34 ± 0.09	1.83 ± 0.25
11	Chieti	Italy (Abruzzo)	(5)	2.3 ± 0.41	0.37 ± 0.12	2.08 ± 0.35
12	Ascoli Piceno	Italy (Marche)	(10)	2.8 ± 0.56	0.37 ± 0.11	1.92 ± 0.30
13	Montecastrilli	Italy (Umbria)	(9)	2.8 ± 0.59	0.42 ± 0.10	2.06 ± 0.29
14	Fabro	Italy (Umbria)	(11)	2.8 ± 0.41	0.42 ± 0.07	1.96 ± 0.21
15	Appennino umbro	Italy (Umbria)	(9)	2.4 ± 0.50	0.36 ± 0.10	1.85 ± 0.27
16	Gubbio	Italy (Umbria)	(25)	3.8 ± 0.75	0.45 ± 0.09	1.97 ± 0.22
17	Città di Castello	Italy (Umbria)	(12)	2.9 ± 0.69	0.39 ± 0.10	1.93 ± 0.28
18	Montemaggiore	Italy (Umbria)	(12)	2.3 ± 0.53	0.29 ± 0.11	1.68 ± 0.27
19	Pietralunga	Italy (Umbria)	(7)	2.4 ± 0.42	0.36 ± 0.10	1.92 ± 0.28
20	Firenze	Italy (Toscana)	(32)	4.1 ± 0.90	0.47 ± 0.08	2.04 ± 0.21
21	Val di Zena	Italy (Emilia-Romagna)	(10)	2.9 ± 0.58	0.43 ± 0.10	2.02 ± 0.27
22	Pianoro	Italy (Emilia-Romagna)	(9)	2.5 ± 0.53	0.38 ± 0.11	1.94 ± 0.20
23	Alba	Italy (Piemonte)	(7)	2.8 ± 0.49	0.42 ± 0.09	2.12 ± 0.29
24	Asti	Italy (Piemonte)	(9)	2.5 ± 0.42	0.40 ± 0.10	1.98 ± 0.27
25	Langhe	Italy (Piemonte)	(15)	3.0 ± 0.76	0.42 ± 0.11	2.01 ± 0.30
26	Pavia	Italy (Lombardia)	(11)	2.4 ± 0.42	0.36 ± 0.10	1.85 ± 0.26
27	Istria	Croatia-Slovenia	(17)	3.0 ± 0.60	0.41 ± 0.10	1.93 ± 0.25
28	Southern Hungary	Hungary	14	3.1 ± 0.74	0.37 ± 0.11	1.91 ± 0.30
29	Šabac	Serbia	6	2.1 ± 0.44	0.28 ± 0.11	1.78 ± 0.31
30	Valjevo	Serbia	7	2.0 ± 0.33	0.33 ± 0.10	1.66 ± 0.27
31	Anina	Romania	3	1.4 ± 0.18	0.17 ± 0.08	ND^{++}
32	Craiova	Romania	22	3.6 ± 0.84	0.44 ± 0.12	2.04 ± 0.31
33	Oryahovo	Bulgaria	6	2.0 ± 0.27	0.39 ± 0.09	1.89 ± 0.22
34	Razgrad	Bulgaria	6	1.9 ± 0.40	0.29 ± 0.11	1.75 ± 0.32
35	Kastoria	Greece	18	3.6 ± 0.71	0.47 ± 0.11	2.13 ± 0.28
36	Elatochori	Greece	6	2.5 ± 0.46	0.40 ± 0.10	2.15 ± 0.33
Mean	Italy	-	12.46 ± 1.52	2.84 ± 0.12	0.40 ± 0.01	1.99 ± 0.03
Mean	Pannonia/Balkans	-	10.50 ± 2.09	2.52 ± 0.24	0.36 ± 0.03	1.92 ± 0.05
Mean	Total	-	11.92 ± 1.23	2.75 ± 0.11	0.39 ± 0.01	1.97 ± 0.02

* Samples derived from the dataset of Rubini et al. [4] are indicated in parentheses. ** Standard errors of the means are reported. ND^{++}: Not determined due to the small sample size and presence of missing data.

2.2. Genetic Diversity Data and Population Structure Analyses

The mean number of alleles (Na) and expected heterozygosity (He) for each locus and for each population were evaluated using GenAlEx v. 6.501 [21]. Allelic richness (Ar) was calculated with the software ADZE [22] using the rarefaction method [23] to correct for differences in sample size. The Ar was weighted to four individuals by excluding calculation for Population 1 and 31, which have a very small sample size. To avoid losing most of the information due to the small sample size of some populations, Ar was also calculated by sorting individuals into eight geographical groups according to their proximity (Figure S1). In this case, it was possible to consider a larger sample size, up to $n = 15$.

To estimate the degree of differentiation among populations, the analysis of molecular variance (AMOVA) and calculation of Fst and Rst values were performed using Arlequin software, version 3.5.1.2 [24]. These analyses were carried out both by comparing all 36 populations (sampling locations) and two regional groups of populations: Italy and Balkans/Pannonia.

Multilocus version 1.3b [25] was used to calculate the number of multilocus genotypes (MLGs) and genotypic diversity (i.e., the probability that two individuals taken at random have different genotypes). To evaluate if samples sharing the same MLG were true clones or resulted from random mating, the Psex (the probability of obtaining the same MLG from different sexual events) values and their significance levels were calculated with MLGsim software [26]. To test for the presence of an isolation by distance pattern, correlation between genetic and geographic distances was evaluated by performing a Mantel test according to Rousset [27] and using the software GenAlEx. The geographic distance matrix, consisting of the natural logarithm of the pairwise distance among populations, was calculated with GenAlEx considering the average geographic coordinates of each population. A genetic distance matrix consisting of pairwise Rst/(1−Rst) values was calculated using the software SPAGeDi version 1.5 [28].

The genetic structure of *T. magnatum* populations was evaluated by Bayesian analysis using the software STRUCTURE version 2.3.4 [29] and TESS version 2.3 [30,31]. Five independent runs of STRUCTURE for K (max number of estimated clusters) ranging from 2 to 10 were conducted. For each K, 2000,000 MCMC (Markov Chain Monte Carlo) and a burn-in of 200,000 iterations were performed, respectively. The admixture and no admixture models were tested considering correlated and uncorrelated allele frequencies. The optimal K was determined by comparing both the log-likelihood values and Evanno's ΔK [32] using the software Structure Harvester [33]. TESS analysis was performed both under the no admixture and the CAR admixture models by conducing 50 runs for each *K* ranging from 2 to 10 with 50,000 total MCMC steps and a burn-in of 10,000 sweeps. The spatial interaction parameter was set to the default value of 0.6 and the degree of trend to linear. To estimate the best K, the Deviance Information Criterion (DIC) was averaged across runs for each *K*. The smallest value before reaching a plateau was selected as the best K. Burn-in length and number of MCMC interactions for STRUCTURE and TESS were established by checking the convergence of summary statistics, and by evaluating consistence among runs of different lengths, following the recommendations in the software manuals.

STRUCTURE and TESS results were processed with the software CLUMPP [34] using the Greedy algorithm, random input order of runs and 1000 repeats. CLUMPP results were used to generate bar graphs using DISTRUCT [35]. The ancestry coefficients calculated with STRUCTURE were also plotted into a geographic map using the "POPSutilities" R script [36] according to the interpolation procedure described by Francois [37]. Interpolate values of ancestry coefficients among each pair of samples were calculated with R using the CLUMPP Q-matrix and a spatial grid obtained from the raster map of the area. For each sample, transition between one group (K) to another was displayed with a progressive change in the color intensity. The starting colors are those assigned to the different K.

3. Results

3.1. Genetic Diversity of T. magnatum Populations

SSR data were generated for 111 *T. magnatum* samples. Most of the samples were from the Balkan/Pannonian region (9 populations, 88 samples) and a few from central (Abruzzo-Molise, 18 samples) and southern Italy (Basilicata and Calabria, 5 samples) (Table S2). SSR analysis always showed the presence of a single allele per locus as expected for haploid organisms. In total, 49 alleles were detected and among them and 12 were new with respect to the alleles previously identified [4] (Table S2). The data obtained in this study were then merged with the data from Rubini et al. [4], which were relative to samples mainly collected in Italy, to end up with a dataset of 429 samples grouped into 36 populations covering almost all of the known *T. magnatum* distributional area (Table 1 and Figure 1).

Figure 1. Map showing the geographical location of the *Tuber magnatum* samples analyzed. Populations are indicated with circles and numbered as in Table 1.

Considering this merged dataset, the number of alleles was 77 with a minimum of 3 alleles for the locus MA13 and a maximum of 19 for the locus MA4. The expected heterozygosity (He) ranged from 0.12 (MA13) to 0.84 (MA51) (Table S3). In each population the mean number of alleles ranged from 1.4 to 4.6 and the expected heterozygosity from 0.17 to 0.56 (Table 1). Comparison of allele distribution between samples from the Italian and Balkan/Pannonian regions revealed 10 private alleles specific to Italy and 28 to the Balkans/Pannonia. Some of these 38 alleles were also private for single populations, but their frequency was no higher than 0.041 (Table S4). The remaining 39 alleles were shared between the two regions, the most of them without relevant differences. Only a few showed a biased frequency, this was the case of the alleles 162 and 172 at MA4 and MA14 loci, respectively, being markedly more frequent among the Balkan/Pannonian samples, and the allele 121 at locus MA21 more frequent among Italian samples (Table S4). By combining the allelic profiles at the eight SSR loci, 362 MLGs were identified. Among these MLGs, 48 were shared at least between two individuals, but only a few (9) turned out to be clones as they had Psex values that were significantly lower, thus rejecting the hypothesis of their origin by sexual reproduction (Table S5). These putative clones were detected only within a population. Some MLGs were also shared among both close and distant populations but none of them showed significant Psex values, suggesting that they were generated by chance because of random mating (Table S5). The overall genotypic diversity was 0.998. When plotted against the number of loci, the genotypic diversity reached a value higher than 0.99 at six loci and also slightly increased up to eight loci (Figure S2).

The Ar for each single population ranged from 1.66 to 2.27. The highest values were observed for Abruzzo-Molise (Populations 4, 5 and 7) among the Italian populations and for Greece (Populations 35 and 36) among the Balkan/Pannonian populations (Table 1). Grouping individuals at a large geographical scale (i.e., 8 groups) allowed us to calculate the Ar by means of a rarefaction analysis based on a larger sample size (i.e., $n = 15$) and to show that the individuals from central-south Italy (Group 2) and those from the southern Balkans (Group 8) had the highest Ar values: 3.50 and 3.66, respectively (Figure S1).

3.2. *The Balkan/Pannonian and Italian Populations Were Genetically Differentiated*

The presence of a genetic structure was revealed by AMOVA, which showed a marked and significant genetic differentiation between the 36 populations (Fst 0.169, $p < 0.001$; Rst = 0.422, $p < 0.001$). A higher significant differentiation (Fst 0.212, $p < 0.001$; Rst = 0.427, $p < 0.001$) was detected when two regional groups of populations (Italian and Balkan/Pannonian) were considered (Table S6). Furthermore, the Mantel test showed that the among-populations differentiation significantly increased with geographic distance ($R^2 = 0.0232$, $p < 0.005$; Figure 2).

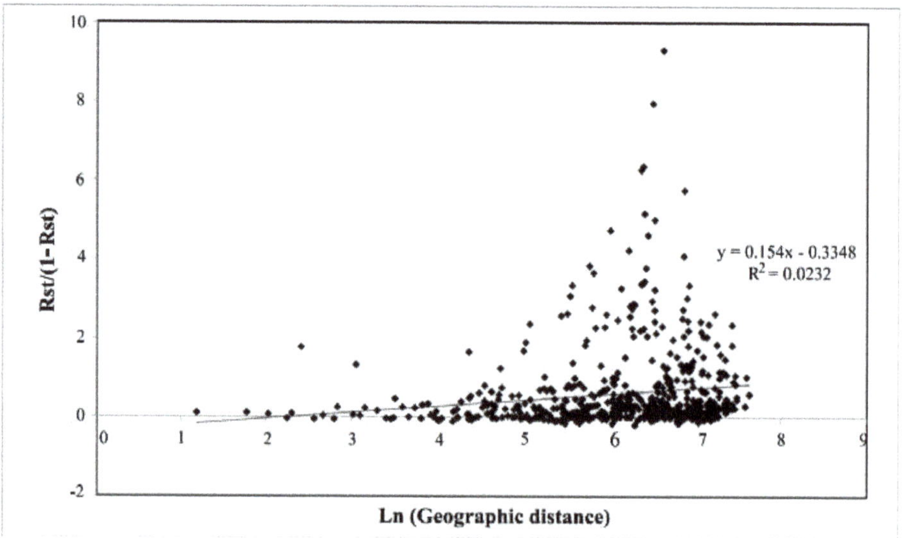

Figure 2. Mantel correlation between the genetic and geographic distances.

To better evaluate the genetic structure of *T. magnatum* populations, an admixture analysis was performed. Bayesian clustering using the software STRUCTURE, showed that the most probable number of clusters (K) was four (Figure S3A,B).

Three of these clusters were primarily associated with different geographical areas of sample acquisition, with one grouping most of the individuals from southern Italy, one those from central-northern Italy and Istria, and the third those from Balkans/Pannonia. Conversely, neither a specific nor a prevalent geographical provenance emerged within individuals of the fourth cluster (Figure 3A). We also noted that some individuals of Balkans/Pannonia shared common ancestry with those of central-northern Italy, Istria and with Population 1 from southern Italy. Plotting the ancestry coefficients obtained with STRUCTURE into the geographical map produced a better picture of the geographical distribution of individuals belonging to the four genetic clusters since they clearly matched the four distinct geographical areas: south Italy, central Italy, central-northern Italy and Istria as well as the Balkans/Pannonia (Figure 3B). No further relevant differentiation among individuals emerged when a higher value of K was considered (Figure 3A). Similar results were obtained using the no-admixture model considering either correlated or uncorrelated allele frequencies (data not shown). When STRUCTURE was run to analyze the Balkan/Pannonian regional group only, a further sub-structuration, with the differentiation of Greek Populations 35 and 36 starting from K = 3, emerged (Figure S4). Conversely, the same analysis on samples from Italy and Istria did not show any further sub-structuration (data not shown).

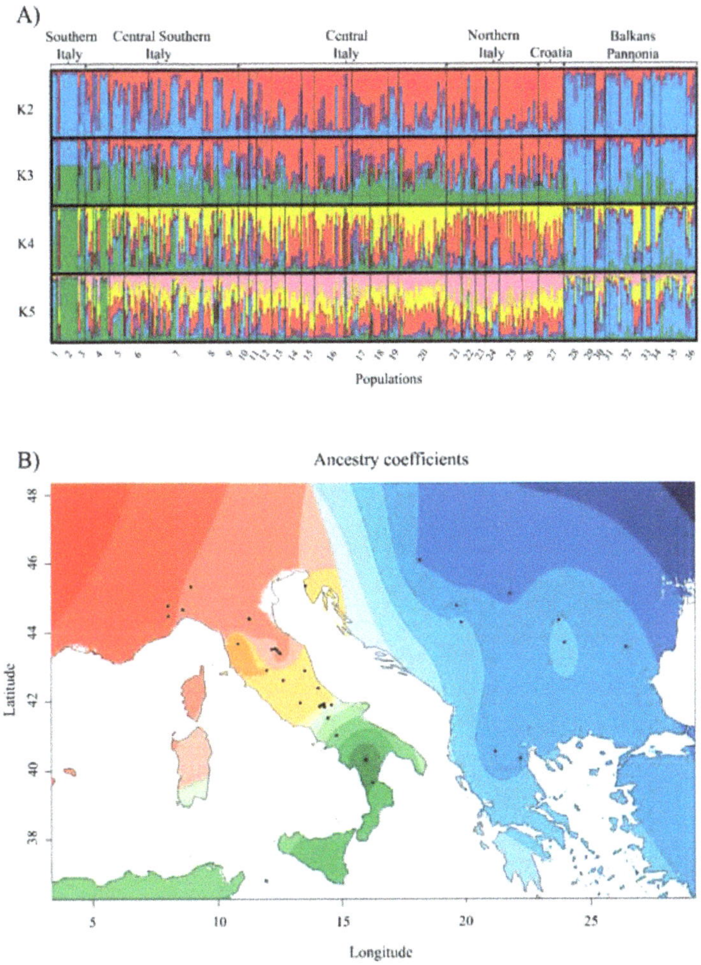

Figure 3. Bayesian analysis of genetic structure. (**A**) Plot of the ancestry coefficients of each single individual obtained with STRUCTURE based on admixture model and correlated allele frequencies. Each K is represented by a different color. Populations are indicated below the figure and their geographic origin above. (**B**) Spatial interpolation of population structure inferred for $K = 4$. Black dots represent the samples locations. The four K groups are represented by green, yellow, red and blue color gradients.

The genetic structure of populations was further explored using TESS. Unlike STRUCTURE, this software estimates the number of genetic clusters (K) by taking into account geographical coordinates of individuals to detect discontinuities in allele frequencies. Running TESS, under the admixture model there were three clusters matching; basically, the three main clusters found by STUCTURE (data not shown). The no-admixture model was more informative since, according to the DIC values, the most probable number of clusters was five (Figure 4A). In agreement with STRUCTURE, TESS evidenced a clear differentiation of southern Italian and Balkan/Pannonian populations. Moreover, the same analysis showed a further differentiation between populations of southern-central Italy and those of central and northern Italy. A fifth group was represented by the Balkan Population 35 from Greece and Population 1 from Calabria, which showed partial common ancestry (Figure 4B).

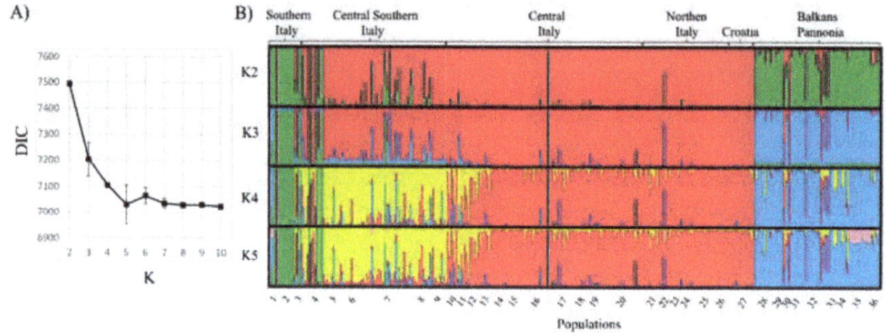

Figure 4. TESS analysis considering a no-admixture model. (**A**) Values of DIC in relation to the values of K. (**B**) Bar plots. Each individual is represented by vertical columns and the proportions of each K are represented by different colors. Populations and their geographic origin are given below and above the figure, respectively.

4. Discussion

When *T. magnatum* is evoked, what comes to mind varies according to people: According to the most gourmets and consumers, *T. magnatum* is the premium truffle, whereas for mycologists it represents the *Tuber* species that, among those of economic relevance, has the narrowest distributional range and the least understood autecology [13]. Previous studies revealed the presence of a genetic structure of *T. magnatum* Italian populations with southernmost and north-westernmost populations genetically differentiated from those of central Italy and Istria [4]. In these studies, only few samples other than those coming from Italy were considered; but, differently from what has been believed in the past, the *T. magnatum* distributional range is not indeed confined to the Italian peninsula only. Rather, it adds to the trans-Adriatic species as an increasing number of works have recently reported on the recovery of this fungus in different Balkan countries [16]. Thus, here we took advantage of *T. magnatum* specimens harvested from these new hotspots of white truffle production, spanning from Hungary, the northernmost, to Bulgaria, the easternmost, and down to Greece, to assess the genetic variability of this species and to ascertain whether genetic differentiation exists between populations of the two main regions of the *T. magnatum* distributional range: Italy and the Balkans/Pannonia. The analysis of more than 400 individuals, distributed into 36 populations located in Italy and the Balkan/Pannonian region and genotyped with eight SSR loci, unveils the presence of a high genetic variability and a clear genetic structure of *T. magnatum* populations.

4.1. New Insights into Genetic Diversity of White Truffle

The study by Rubini and colleagues [4] conducted over more than 300 *T. magnatum* samples harvested in Italy and Istria and genotyped by means of a few SSR markers was the first to show the presence of genetically structured populations and the occurrence of an extensive gene flow within and among *T. magnatum* populations, likely thanks to spore dispersal by mycophagist mammals. With respect to previous studies, not only the higher sample size but also the use of the more informative markers, that is, polymorphic SSRs vs. isoenzymes, RAPD or SNPs on highly conserved loci (i.e., β tubulin), were therefore crucial to these authors to gain first evidence on the abundance and distribution of the genetic variability in this truffle species. Yet, the use of a handful of polymorphic SSR loci coupled to a sampling of *T. melanosporum* all over the distributional range was sufficient to Riccioni et al. [5] to confute the thesis of a trifling genetic polymorphism in this species [38] and map its possible post glacial recolonization pattern. According to this observation, here we tested the hypothesis that even a relatively small number of polymorphic SSR loci would be sufficient to shed light into *T. magnatum* genetic diversity, if samples representative of the entire species distributional range are analyzed. Following this rationale, the first goal of the present study was to broaden the *T. magnatum* sampling areas. Thus, despite the widely known and hard to counteract secrecy of truffle hunters, an extensive sampling of *T. magnatum* fruitbodies from Bulgaria, Hungary, Romania, Serbia and Greece has been performed. Eighty-eight samples from these countries and a few samples from south Italy have been genotyped for eight SSR loci and these data merged with the SSR profiles retrieved from Rubini et al. [4]. By doing so, the number of alleles detected among the 429 specimens increased up to 77, with 12 new alleles being found, and with the number of alleles per locus (from 3 up to 19) in the same range found for other truffle species, such as *T. melanosporum* (2–18) and *T aestivum* (4–15) [39,40].

The value of the He over loci of 0.54 (Table S4) or 0.38, if the average value among populations is considered (Table 1), is much higher than that reported for this species (0.16) when sampling was confined to southern Italy only [41]. This high value of He suggests that at large geographical scale the level of genetic diversity in *T. magnatum* is comparable with that of other European species with a wider geographical distribution, such as *T. melanosporum* and *T. aestivum,* which showed He values of 0.41 and 0.50, respectively [7,39]. The high level of genetic diversity is also confirmed by the high number of MLGs: 332 over 429 fruit bodies analyzed and the overall genotypic diversity of 0.99, a value similar to that found in *T. melanosporum* and *T. aestivum* [7,39]. Moreover, the evidence that genotypic diversity reaches a plateau value of 0.99 with just six loci indicates that the number of loci used in this study is adequate to evaluate the genetic variability of *T. magnatum*.

If the entire sample set is split into two main regions, the Italian and the Balkan/Pannonian ones, then the presence of many private alleles specific to a single region emerges, although with a very low frequency. It is worth mentioning that a few of the shared alleles show a quite different frequency between the two regions. In sum, enlarging the sample size with individuals from the easternmost and southernmost *T. magnatum* distributional areas has allowed us to disclose new and private alleles and gain a closer look into the population genetics of this species.

4.2. Large Scale Sampling Reveals a Phylogeographic Structure in T. magnatum

The AMOVA analysis of the SSR data obtained from the 429 samples proves the presence of a genetic structure among populations. The values of Fst and Rst, in fact, are higher in the present research vs. the previous work by Rubini et al. [4], suggesting that the new populations considered, mainly from the Balkans/Pannonia, are genetically differentiated from those of Italy and Istria. This is confirmed by the Fst and Rst values that not only are significant but also increase when two regional groups of populations (Balkan/Pannonia and Italy) are compared. In keeping with this, the Mantel test depicts an isolation by distance pattern (Figure 2).

The presence of a genetic structure has been evaluated more in detail by performing Bayesian analysis. The STRUCTURE algorithm clearly reveals that Balkan/Pannonian and Italian populations are genetically differentiated. Moreover, Italian populations are split into tree genetic clusters: southern, central and central-northern Italy, in agreement with results of Rubini et al. [4].

Historical population expansions and restrictions resulting from climate changes have been reported for plant species, including truffle host species, which experienced population bottlenecks as a consequence of glaciations [42]. In concert with this, the geographical distribution of European truffle species has followed the population expansion and restriction processes of their hosts [38,43]. For example, within the black truffle clade, *T. melanosporum* survived in refugia located in the Iberian and Italian peninsulas, as inferred by ITS, ISSR and SSR markers [3,5,44]. Conversely the surviving pattern of *T. brumale* aggr., resulting from the phylogeny of ITS, LSU and PKC loci, was more complex: Within the *T. brumale* clade A, populations of haplotype I survived the last glaciation in Western Europe, those of haploytpe II in Eastern Europe whereas those of clade B in the Carpathian basin and Balkan region [6]. This latter clade was later proposed as representing a cryptic species, *T. cryptobrumale* [45]. Concerning the *T. aestivum* clade, according to the distribution and relatedness of the ITS haplotypes of samples all over Europe and Turkey, it has been suggested that this species survived in Turkey whereas European populations likely experienced a population bottleneck during the last glaciation [9]. The geographic structure of *T. magnatum* populations from the Italian peninsula, as per SSR analyses, was consistent with the occurrence a glacial refugium in central Italy from which the northernmost and southernmost populations originated [4]. Thanks to the large sampling performed, this hypothesis is here reinforced as among Italian populations, those from the central-southern area exhibit the highest levels of allelic richness. Our study also suggests that the Balkan peninsula may have represented a *T. magnatum* glacial refugium as well. Our inference stems from the following considerations: (i) Balkan/Pannonian populations, with a few exceptions, belong to a different genetic cluster with respect to Italian specimens, suggesting an independent evolutionary history; (ii) STRUCTURE analysis performed in Balkan/Pannonian populations shows that the individuals from the southernmost populations (Greece) tend to differentiate from the others and the He allelic richness is higher in these populations, a situation frequently expected in correspondence to putative glacial refugia [46]; and (iii) many truffle host plant species, including those that host *T. magnatum* (i.e., beech and hornbeam spp.) survived the last glaciation in three main Mediterranean peninsulas: the Iberian, Italian and Balkan ones [47–49].

It is noteworthy that, although a general phylogeographic pattern emerged from the STRUCTURE analysis, some individual from the Balkan/Pannonian region (e.g., individuals from Populations 30, 32 and 34) share common ancestry with individual from Italian populations. Moreover, analyses that take in account the geographical information (i.e., TESS algorithm), confirm clustering into four groups but also show that the individuals from Population 1 (southern Italy) share partial ancestry with those from Population 35, one of the southernmost populations from the Balkan area.

The finding that individuals across the two shores of the Adriatic sea partially share a common ancestry poses the question whether strain migration occurred from one side of the Adriatic shore to the other. In fact, the hypothesis that, in the past, strains from southern-central Italy moved to the Balkan region or vice versa, cannot be ruled out. Many Mediterranean taxa present disjunct distributions between the west and east Mediterranean, and these disjunct biogeographical patterns are the results of the complex paleogeographic history of the present Mediterranean region [50]. Land bridges between the Italian and Balkan shores of the Adriatic sea occurred in the Neogene through the formation of the Apulo-Dalmatic Realm [51] and likely during the Pleistocene glaciations [52]. Thus, it is conceivable that by enabling the migration of mycophagist animals these geological events may have favored truffle spore dispersal between the two peninsulas. Along the same reasoning, as the two shores of the Adriatic sea shared basically the same climate and soil (calcareous soil of cretaceous origin) conditions [53], it is more than conceivable that whatever the truffle migration direction was, the newly introduced truffle strains encountered host species and pedoclimatic conditions that have favored their settlement. Mycophagist-mediated spore dispersal across the Alps also appears to be the most conceivable explanation of the higher relatedness of Istrian specimens to those from Italy rather than to those from the Balkans. A more extensive sampling of specimens from both sides of the Adriatic sea covering, in particular, the latitudes spanning from 40° to 42° N, coupled with the use of phylogenetically informative functional markers, would help us to further test the "bridge" hypothesis.

4.3. Genetic Structure and Conservation Implications

Unearthing *T. magnatum* population genetic structure may have important implications for conservation of its biodiversity. Differently from black truffle species, *T. magnatum* cultivation is not yet established [13]; thus, propagation of this species is almost exclusively natural. In vitro isolation of mycelium strains is also very challenging as this species shows a very slow growth rate and optimal nutritional requirements have not been identified yet. Thus, the preservation of *T. magnatum* strains ex situ in genetic banks is currently unfeasible. Rather, the most affordable strategy for the conservation of *T. magnatum* biodiversity relies on the preservation of its natural habitats. On these premises, results of the present study may be of relevance to identify and preserve populations and strains specific and adapted to different environments.

5. Conclusions

Here we have shown that *T. magnatum* genetic diversity is higher than hitherto thought and geographically structured across the Italian peninsula and Balkan/Pannonia region. Our findings are of relevance to make inferences about the phylogeographic history of this species but also for marketing and conservation purposes. The price of white truffles is traditionally dictated by their geographic provenance; thus, by increasing the number of genetic markers it would be possible, in the near future, to trace the origin of these truffles. This is a prerequisite to promote and sustain local white truffle-linked ecosystem services and economies but also to evaluate if and to what extent genetic determinants concur to shape the aroma variability across white truffles of different provenance. From a conservation point-of-view, the presence of a phylogeographic structure led us to hypothesize that *T. magnatum* strains of different geographic areas might exhibit different adaptation traits. In the light of the difficulties in its cultivation/propagation and of a global warming scenario, preservation of *T. magnatum* natural habitats from both Italy and the Balkan/Pannonian countries is therefore crucial to prevent the erosion of its biodiversity.

Supplementary Materials: The following are available online at http://www.mdpi.com/1424-2818/12/2/44/s1, Table S1: Panels of SSR loci used, Table S2: List of the samples considered in this study, Table S3: Polymorphism levels of the 8 SSRs over the entire sample set, Table S4: Allele distribution and frequency in samples from Italian (I) and Balkan/Pannonian regions (B), Table S5: List of MLG and results of MLGsim analysis. Populations are indicated when identical MLG are found, Table S6: AMOVA analysis among all populations and considering two regional groups of populations (Italy, and Balkans/Pannonia), Figure S1: Allelic richness in eight geographical groups. The geographical groups are indicated below the figure. Each geographical group includes all individuals from populations (numbered as in Table 1) reported in parentheses, Figure S2: Average genotypic diversity in function of the number of loci, Figure S3: Estimation of the most probable k. (a) Mean log likelihood over 5 runs (error bars = standard deviations) and (b) ΔK, the second order rate of change in the likelihood at each K, Figure S4: STRUCTURE analysis performed on Balcan/Pannonian populations only, based on admixture model and correlated allele frequencies. Each K is represented by a different color. Populations are indicated below the figure and their geographic origin above.

Author Contributions: Conceptualization, A.R., G.P. and F.P.; methodology, A.R., B.B. and C.R.; investigation, data curation, B.B., V.D., M.L. and C.R.; formal analysis, A.R. and B.B.; resources, G.P. and M.L.; writing—original draft preparation, A.R., G.P., F.P.; writing—review and editing, B.B., V.D., C.R., M.L., G.P., F.P. and A.R. All authors have read and agree to the published version of the manuscript.

Funding: This research received no external funding.

Acknowledgments: We are grateful to Pavlina Kladopoulou, Giorgio Konstanidis, Kiro Prodan, Kenan Kyose, Iordan Taralanski, Dejan Polic, Oszkar Fekete, Istvan Bagi, Silvio Guardiani, Gianni Miglietta, Mario Marchione, Marilena Oddis, Simona Ascione, Domenico Puntillo for providing us with fresh specimens of *Tuber magnatum* from the Balkans and Hungary and some areas of the Italian peninsula.

Conflicts of Interest: The authors declare no conflict of interest.

References

1. Smith, S.E.; Read, D.J. *Mycorrhizal Symbioses*, 3rd ed.; Academic Press: Cambridge, MA, USA, 2008.
2. Trappe, J.M.; Molina, R.; Luoma, D.L.; Cázares, E.; Pilz, D.; Smith, J.E.; Castellano, M.A.; Miller, S.L.; Trappe, M.J. *Diversity, Ecology and Conservation of Truffle Fungi in Forests of the Pacific Northwest*; Gen. Tech. Rep. PNW-GTR-772; USDA Forest Service, Pacific Northwest Research Station: Portland, OR, USA, 2009.
3. Murat, C.; Díez, J.; Luis, P.; Delaruelle, C.; Dupré, C.; Chevalier, G.; Bonfante, P.; Martin, F. Polymorphism at the ribosomal DNA ITS and its relation to postglacial re-colonization routes of the Perigord truffle *Tuber melanosporum*. *New Phytol.* **2004**, *164*, 401–411. [CrossRef]
4. Rubini, A.; Paolocci, F.; Riccioni, C.; Vendramin, G.G.; Arcioni, S. Genetic and phylogeographic structures of the symbiotic fungus *Tuber magnatum*. *Appl. Environ. Microbiol.* **2005**, *71*, 6584–6589. [CrossRef] [PubMed]
5. Riccioni, C.; Belfiori, B.; Rubini, A.; Passeri, V.; Arcioni, S.; Paolocci, F. *Tuber melanosporum* outcrosses: Analysis of the genetic diversity within and among its natural populations under this new scenario. *New Phytol.* **2008**, *180*, 466–478. [CrossRef] [PubMed]
6. Merényi, Z.; Varga, T.; Geml, J.; Orczán, Á.K.; Chevalier, G.; Bratek, Z. Phylogeny and phylogeography of the *Tuber brumale* aggr. *Mycorrhiza* **2014**, *24*, 101–113. [CrossRef] [PubMed]
7. Molinier, V.; Murat, C.; Peter, M.; Gollotte, A.; De la Varga, H.; Meier, B.; Egli, S.; Belfiori, B.; Paolocci, F.; Wipf, D. SSR-based identification of genetic groups within European populations of *Tuber aestivum* Vittad. *Mycorrhiza* **2016**, *26*, 99–110. [CrossRef] [PubMed]
8. Qiao, P.; Tian, W.; Liu, P.; Yu, F.; Chen, J.; Deng, X.; Wan, S.; Wang, R.; Wang, Y.; Guo, H. Phylogeography and population genetic analyses reveal the speciation of the *Tuber indicum* complex. *Fungal Genet. Biol.* **2018**, *113*, 14–23. [CrossRef]
9. Riccioni, C.; Rubini, A.; Türkoğlu, A.; Belfiori, B.; Paolocci, F. Ribosomal DNA polymorphisms reveal genetic structure and a phylogeographic pattern in the Burgundy truffle *Tuber aestivum* Vittad. *Mycologia* **2019**, *111*, 26–39. [CrossRef]
10. Gandeboeuf, D.; Dupré, C.; Roeckel-Drévet, P.; Nicolas, P.; Chevalier, G. Grouping and identification of *Tuber* species using RAPD markers. *Can. J. Bot.* **1997**, *75*, 36–45. [CrossRef]
11. Mello, A.; Fontana, A.; Meotto, F.; Comandini, O.; Bonfante, P. Molecular and morphological characterization of *T. magnatum* mycorrhizas in a long-term survey. *Microbiol. Res.* **2001**, *155*, 279–284. [CrossRef]
12. Mello, A.; Murat, C.; Vizzini, A.; Gavazza, V.; Bonfante, P. *Tuber magnatum* Pico, a species of limited geographical distribution: Its genetic diversity inside and outside a truffle ground. *Environ. Microbiol.* **2005**, *7*, 55–65. [CrossRef]
13. Riccioni, C.; Rubini, A.; Belfiori, B.; Gregori, G.; Paolocci, F. Tuber magnatum: The Special One. What Makes It so Different from the Other *Tuber* spp. In *True Truffle (Tuber spp.) in the World: Soil Ecology, Systematics and Biochemistry*; Zambonelli, A., Iotti, M., Murat, C., Eds.; Springer: Cham, Switzerland, 2016; pp. 87–103. [CrossRef]
14. Frizzi, G.; Lalli, G.; Miranda, M.; Pacioni, G. Intraspecific isozyme variability in Italian populations of the white truffle *Tuber magnatum*. *Mycol. Res.* **2001**, *105*, 365–369. [CrossRef]
15. Paolocci, F.; Rubini, A.; Riccioni, C.; Arcioni, S. Reevaluation of the life cycle of *Tuber magnatum*. *Appl. Environ. Microbiol.* **2006**, *72*, 2390–2393. [CrossRef] [PubMed]
16. Marjanović, Ž.; Grebenc, T.; Marković, M.; Glišić, A.; Milenković, M. Ecological specificities and molecular diversity of truffles (genus Tuber) originating from mid-west of the Balkan Peninsula. *Sydowia* **2010**, *62*, 67–87.
17. Tabouret, P. Exclusif! Description d'un site français producteur de truffe blanche d'Italie. *Le Trufficulteur* **2011**, *78*, 18.
18. Tabouret, P. Constitution d'une fédération Suisse de la truffe et de la trufficulture. *Le Trufficulteur* **2012**, *81*, 25.
19. Paolocci, F.; Rubini, A.; Granetti, B.; Arcioni, S. Rapid molecular approach for a reliable identification of *Tuber* spp. ectomycorrhizae. *FEMS Microbiol. Ecol.* **1999**, *28*, 23–30. [CrossRef]
20. Rubini, A.; Topini, F.; Riccioni, C.; Paolocci, F.; Arcioni, S. Isolation and characterization of polymorphic microsatellite loci in white truffle (*Tuber magnatum*). *Mol. Ecol. Notes* **2004**, *4*, 116–118. [CrossRef]
21. Peakall, R.; Smouse, P.E. GenAlEx 6.5: Genetic analysis in Excel. Population genetic software for teaching and research—An update. *Bioinformatics* **2012**, *28*, 2537–2539. [CrossRef]

22. Szpiech, Z.A.; Jakobsson, M.; Rosenberg, N.A. ADZE: A rarefaction approach for counting alleles private to combinations of populations. *Bioinformatics* **2008**, *24*, 2498–2504. [CrossRef]
23. El Mousadik, A.; Petit, R.J. High level of genetic differentiation for allelic richness among populations of the argan tree [*Argania spinosa* (L.) Skeels] endemic to Morocco. *Theor. Appl. Genet.* **1996**, *92*, 832–839. [CrossRef]
24. Excoffier, L.; Lischer, H.E. Arlequin suite ver 3.5: A new series of programs to perform population genetics analyses under Linux and Windows. *Mol. Ecol. Resour.* **2010**, *10*, 564–567. [CrossRef] [PubMed]
25. Agapow, P.M.; Burt, A. Indices of multilocus linkage disequilibrium. *Mol. Ecol. Notes* **2011**, *1*, 101–102. [CrossRef]
26. Stenberg, P.; Lundmark, M.; Saura, A. MLGsim: A program for detecting clones using a simulation approach. *Mol. Ecol. Resour.* **2003**, *3*, 329–331. [CrossRef]
27. Rousset, F. Genetic differentiation and estimation of gene flow from F-statistics under isolation by distance. *Genetics* **1997**, *145*, 1219–1228. [PubMed]
28. Hardy, O.J.; Vekemans, X. SPAGEDI: A versatile computer program to analyze spatial genetic structure at the individual or population level. *Mol. Ecol. Notes* **2002**, *2*, 618–620. [CrossRef]
29. Pritchard, J.K.; Stephens, M.; Donnelly, P. Inference of population structure using multilocus genotype data. *Genetics* **2000**, *155*, 945–959.
30. Chen, C.; Durand, E.; Forbes, F.; Francois, O. Bayesian clustering algorithms ascertaining spatial population structure: A new computer program and a comparison study. *Mol. Ecol. Notes* **2007**, *7*, 747–756. [CrossRef]
31. Durand, E.; Jay, F.; Gaggiotti, O.E.; François, O. Spatial inference of admixture proportions and secondary contact zones. *Mol. Biol. Evol.* **2009**, *26*, 1963–1973. [CrossRef]
32. Evanno, G.; Regnaut, S.; Goudet, J. Detecting the number of clusters of individuals using the software STRUCTURE: A simulation study. *Mol. Ecol.* **2005**, *14*, 2611–2620. [CrossRef]
33. Dent, A.E.; von Holdt, B.M. STRUCTURE HARVESTER: A website and program for visualizing STRUCTURE output and implementing the Evanno method. *Conserv. Genet. Resour.* **2012**, *4*, 359–361.
34. Jakobsson, M.; Rosenberg, N.A. CLUMPP: A cluster matching and permutation program for dealing with label switching and multimodality in analysis of population structure. *Bioinformatics* **2007**, *23*, 1801–1806. [CrossRef] [PubMed]
35. Rosenberg, N.A. DISTRUCT: A program for the graphical display of population structure. *Mol. Ecol. Notes* **2004**, *4*, 137–138. [CrossRef]
36. Jay, F.; Manel, S.; Alvarez, N.; Durand, E.Y.; Thuiller, W.; Holderegger, R.; Taberlet, P.; François, O. Forecasting changes in population genetic structure of alpine plants in response to global warming. *Mol. Ecol.* **2012**, *21*, 2354–2368. [CrossRef] [PubMed]
37. How to Display Ancestry Coefficients (Q-matrix) on a Geographic Map. Available online: http://membres-timc.imag.fr/Olivier.Francois/TESS_Plot.html (accessed on 4 November 2019).
38. Bertault, G.; Raymond, M.; Berthomieu, A.; Callot, G.; Fernandez, D. Trifling variation in truffles. *Nature* **1998**, *394*, 734. [CrossRef]
39. Murat, C.; Riccioni, C.; Belfiori, B.; Cichocki, N.; Labbé, J.; Morin, E.; Tisserant, E.; Paolocci, F.; Rubini, A.; Martin, F. Distribution and localization of microsatellites in the Perigord black truffle genome and identification of new molecular markers. *Fungal Genet. Biol.* **2011**, *48*, 592–601. [CrossRef]
40. Molinier, V.; Murat, C.; Morin, E.; Gollotte, A.; Wipf, D.; Martin, F. First identification of polymorphic microsatellite markers in the Burgundy truffle, *Tuber aestivum* (Tuberaceae). *Appl. Plant. Sci.* **2013**, *1*, 1200220. [CrossRef]
41. Pomarico, M.; Figliuolo, G.; Rana, G.L. *Tuber* spp. Biodiversity in one of the southernmost European distribution areas. *Biodivers. Conserv.* **2007**, *16*, 3447–3461. [CrossRef]
42. Petit, R.J.; Brewer, S.; Bordács, S.; Burg, K.; Cheddadi, R.; Coart, E.; Cottrel, J.; Csaikl, U.M.; Van Dam, B.; Deans, J.D.; et al. Identification of refugia and post-glacial colonisation routes of European white oaks based on chloroplast DNA and fossil pollen evidence. *For Ecol. Manag.* **2002**, *156*, 49–74. [CrossRef]
43. Bertault, G.; Rousset, F.; Fernandez, D.; Berthomieu, A.; Hochberg, M.E.; Callot, G.; Raymond, M. Population genetics and dynamics of the black truffle in a man-made truffle field. *Heredity* **2001**, *86*, 451–458. [CrossRef]
44. García-Cunchillos, I.; Sánchez, S.; Barriuso, J.J.; Pérez-Collazos, E. Population genetics of the westernmost distribution of the glaciations-surviving black truffle *Tuber melanosporum*. *Mycorrhiza* **2014**, *24*, 89–100. [CrossRef]

45. Merényi, Z.; Varga, T.; Hubai, A.G.; Pitlik, P.; Erős, Á.; Trappe, J.M.; Bratek, Z. Challenges in the delimitation of morphologically similar species: A case study of *Tuber brumale* agg. (Ascomycota, Pezizales). *Mycol. Prog.* **2017**, *16*, 613. [CrossRef]
46. Widmer, A.; Lexer, C. Glacial refugia: Sanctuaries for allelic richness, but not for gene diversity. *Trends Ecol. Evol.* **2001**, *16*, 267–269. [CrossRef]
47. Taberlet, P.; Fumagalli, L.; Wust-Saucy, A.G.; Cosson, J.F. Comparative phylogeography and postglacial colonisation routes in Europe. *Mol. Ecol.* **1998**, *7*, 453–464. [CrossRef] [PubMed]
48. Grivet, D.; Petit, R.J. Chloroplast DNA phylogeography of the hornbeam in Europe: Evidence for a bottleneck at the outset of postglacial colonization. *Conserv. Genet.* **2003**, *4*, 47–56. [CrossRef]
49. Magri, D. Patterns of post-glacial spread and the extent of glacial refugia of European beech (*Fagus sylvatica*). *J. Biogeogr.* **2008**, *35*, 450–463. [CrossRef]
50. Sanmartín, I. Dispersal vs vicariance in the mediterranean: Historical biogeography of the Paleartic Pachydeminae (Coleoptera, Scarabeodidea). *J. Biogeog.* **2003**, *30*, 1883–1897. [CrossRef]
51. De Giuli, C.; Masini, F.; Torre, D. Island endemism in the Eastern Mediterranean mammalian paleofaunas patterns in the Gargano paleo-arcipelago. *Atti Accad. Naz. Lin.* **1990**, *85*, 247–262.
52. Kaiser, K. The climate of Europe during the Quaternary Ice Age. In *Quaternary Geology and Climate*; Wright, H.E., Jr., Ed.; National Academy of Sciences: Washington, DC, USA, 1969.
53. Gridelli, E. IL problema delle specie a diffusione attuale transadriatica con particolare riguardo ai Coleotteri. *Ital. J. Zool.* **1950**, *17*, 421–441.

diversity

MDPI

Article

Ecology, Phylogeny, and Potential Nutritional and Medicinal Value of a Rare White "Maitake" Collected in a Mediterranean Forest

Maria Letizia Gargano [1], Georgios I. Zervakis [2], Omoanghe S. Isikhuemhen [3], Giuseppe Venturella [4], Roberta Calvo [4,*], Anna Giammanco [5], Teresa Fasciana [5] and Valeria Ferraro [4]

1 Department of Agricultural and Environmental Science, University of Bari Aldo Moro, Via Amendola 165/A, I-70126 Bari, Italy; marialetizia.gargano@uniba.it

2 Laboratory of General and Agricultural Microbiology, Agricultural University of Athens, 11855 Athens, Greece; zervakis@aua.gr

3 Mushroom Biology and Fungal Biotechnology Laboratory, Department of Natural Resource & Environmental Design, North Carolina A&T State University, Greensboro, NC 27411, USA; omon.isi@fungibiotech.com

4 Department of Agricultural, Food and Forest Sciences, University of Palermo, Viale delle Scienze, Bldg. 5, I-90128 Palermo, Italy; giuseppe.venturella@unipa.it (G.V.); valeria.ferraro@unipa.it (V.F.)

5 Department of Health Promotion, Mother and Child Care, Internal Medicine and Medical Specialities, University of Palermo, 90127 Palermo, Italy; anna.giammanco@unipa.it (A.G.); teresa.fasciana@unipa.it (T.F.)

* Correspondence: roberta.calvo@unipa.it; Tel.: +39-349-090-6252

Received: 4 May 2020; Accepted: 4 June 2020; Published: 8 June 2020

Abstract: Albino *Grifola frondosa* (Dicks.) Gray "maitake" mushrooms (described as *G. albicans* Imazeki and then placed in synonymy with *G. frondosa*) are particularly rare, and the few pertinent records are not treated in scientific publications. A field investigation carried out in Sicily (Italy) led to the collection of an unusual white *Grifola* specimen at the base of a living tree of *Quercus pubescens* Willd. s.l. The outcome of sequencing the internal transcribed spacer (ITS) region of nuclear ribosomal DNA (nrDNA) indicated that it belongs to *G. frondosa* and provided an insight to the phylogenetic relationships within the genus. The results of nutritional composition analysis showed that the albino basidioma possesses relatively high contents of Ca, Fe, K, and Cu and is rather low in Na when compared with literature data on edible mushrooms. Vitamin (B_1, B_2, B_3, B_5, B_9, and D_2) contents ranged from 0.15 to 3.89 mg per 100 g of mushroom dry weight. The cold-water extract of this specimen was effective at inhibiting the growth of *Staphylococcus epidermidis* ATCC 12228 and *Pseudomonas aeruginosa* ATCC 15442 at the maximum screening concentration of 50% *v/v*. In addition, the extract slowed down the ability of *Staphylococcus aureus* ATCC 43300 to form biofilms. According to data hereby reported, the albino *G. frondosa* is a culinary-medicinal mushroom with a promising exploitation potential.

Keywords: *Grifola frondosa*; fungal diversity; Mediterranean forest; medicinal mushroom; bioprospecting; ITS rDNA; phylogenetics; basidiomycete; polypore fungus; *Quercus pubescens*

1. Introduction

The family Grifolaceae Jülich (Polyporales, Basidiomycota) comprises only the genus *Grifola* Gray [1], which consists of nine species according to Index Fungorum. *Grifola frondosa* (Dicks) Gray includes several forms and varieties, such as *G. frondosa* f. *frondosa* (Dicks.) Gray, *G. frondosa* f. *intybacea* (Fr.) Pilát, *G. frondosa* var. *frondosa* (Dicks.) Gray, and *G. frondosa* var. *intybacea* (Fr.) Cetto, while *G. intybacea* (Fr.) Imazeki is considered as a synonym. The culinary-medicinal mushroom

G. frondosa, widely known as "maitake", is a white-rot polypore associated primarily with deciduous trees of different genera (*Quercus* L., *Acer* L., *Carpinus* L., *Castanea* Mill., *Fagus* L., *Ulmus* L.) while it appears more rarely on conifers. This species is distributed in temperate regions of North America, Europe, and Asia (Japan and China) [2]. Albino maitake mushrooms (described as *G. albicans* by Imazeki in 1943 [3] and then placed in synonymy with *G. frondosa*) are quite rare and the few available pertinent records do not appear in scientific publications. Only recently, Kawaguchi et al. [4,5] reported on melanin biosynthesis in *G. frondosa* and suggested that the albino mutation is caused by a single base deletion in the coding region of the tyrosinase 2 (*tyr2*) gene. It is noteworthy that white strains of *G. frondosa* are much sought after since they can be used for culinary purposes without the unwanted dark brown pigment resulting from processing the common form of this mushroom [4].

Edible mushrooms are widely acknowledged for their nutritional and medicinal properties [6]; however, their potential market is still far from being fully developed considering that there is a high consumer demand. Consequently, the private sector is eager to generate new relevant products. In the frame of a research project aiming at collecting mushrooms with bioprospecting potential, a field investigation carried out in Sicily (Italy) led to the collection of an unusual *Grifola* basidioma of white color and considerable weight (ca. 7 kg). The objective of the present work was to provide morphological, molecular/phylogenetic, and ecological data about this rare albino *Grifola* specimen. Information on the nutritional value and antimicrobial activity is also reported.

2. Materials and Methods

2.1. Sample Collection, Habitat Details, and Evaluation of Morphological Characters

Research carried out in forest ecosystems in Sicily (Italy) led to the collection of a single albino *Grifola* basidioma found at the base of a monumental downy oak tree (*Quercus pubescens* Willd. s.l.), 10 January 2016, Castelbuono, Madonie Regional Park (province of Palermo, Sicily), 800 m a.s.l., 37°54′48″ N, 14°04′41″ E. In Castelbuono, summers are rather brief, hot, and dry, with clear skies and condensation often appearing in the late afternoon, while winters are long, cold, rainy, and windy. Annual temperature ranges from 7 to 29 °C (rarely below 4 °C or above 32 °C) while the average value is 15.8 °C. The average annual rainfall is 515 mm. In January, the month when the specimen under study was harvested, the average temperature is 8.9 °C, and the respective maximum and minimum values are 11.5 and 6.3 °C; the rainfall is 67 mm. The forest is characterized by evergreen and deciduous oak vegetation with a prevalence of *Quercus ilex* L. and *Q. pubescens* Willd. s.l. In the upper part of the mountain, the vegetation is characterized by mixed woods of oaks, *Fagus sylvatica* L. and *Ilex aquifolium* L. The specimen was transferred to the Department of Agricultural, Food, and Forest Sciences (SAAF) of the University of Palermo, and stored at <4 °C for up to 24 h prior to morphological examination, which was carried out according to Bernicchia [7]. Observations on macromorphological characters (pileus, cuticle, pores, stipe, context, etc.) were performed on fresh material, while microscopic characters (hyphal system, generative hyphae, basidia, sterigmata, cystidia, cystidioles, and basidiospores) were studied by using 3% potassium hydroxide and ammoniacal Red Congo. In addition, the reaction of the flesh to iodine-potassium iodide, potassium hydroxide, and iron salts was also evaluated.

2.2. Establishment of Pure Cultures and Mushroom Cultivation Trials

The collected basidioma was dried and then deposited in the Herbarium of the Department of Agricultural, Food, and Forest Sciences (SAAF 450). Prior to this, a piece of pseudo-tissue was removed from the fresh specimen, placed on potato dextrose agar (PDA) in Petri dishes under aseptic conditions, and incubated for 15 days at 25 ± 2 °C. The established pure culture was stored in the Mycotheca of the Herbarium SAF (SAF 323), and it was subsequently used for the inoculation of mushroom cultivation substrates composed of either holm oak (*Quercus ilex* L.) or chestnut (*Castanea sativa* Miller) wood residues. The former (holm oak wood shavings) was filled in plastic bottles (volume: 1 L each), while the latter (chestnut wood chips of ca. 2–3 cm and sawdust) was filled in plastic bags (volume: 6 L each) after their moisture content was adjusted to 50–60%; in both cases, non-hydrophilic cotton filters were used to allow air exchange. Sterilization was performed twice at 120 °C, 1.1 Atm for 20 min with a 24-h interval between each cycle. After the final cooling, substrates were inoculated using actively growing mycelium and incubation was carried out at 25 °C in the dark.

2.3. DNA Extraction, PCR Amplification, and Sequencing

Total genomic DNA was extracted from the herbarium specimen using the Nucleospin Plant II DNA kit (Macherey and Nagel, Düren, Germany) following the manufacturer's protocol. The internal transcribed spacer (ITS) region of nuclear ribosomal DNA (nrDNA) was amplified using the primer combination ITS1/ITS4 [8]. Polymerase chain reactions (PCRs) were performed in 50 μL containing 50 ng DNA template, 0.25 μM of each primer, 0.2 mM of each dNTP, 1 × HiFi Buffer (Takara BIO INC., Shiga, Japan), and 1 U HiFi Taq DNA polymerase (Takara BIO INC., Shiga, Japan). Conditions for PCR amplification were as follows: 94 °C for 5 min, followed by 35 cycles of 94 °C for 30 s, 50 °C for 30 s and 72 °C for 1 min, and a final extension at 72 °C for 10 min.

PCR products were purified using an Invitrogen PureLink kit (Thermo Fisher Scientific, Seoul, Korea) and were submitted for sequencing to CeMIA S.A. (Larissa, Greece). The resulting chromatograms were proofread, and the sequence generated was deposited in GenBank under the accession number MN944407.

2.4. Phylogenetic Analysis of Sequence Data

In addition to the biological material examined in this study, 53 additional ITS sequences were included in the phylogenetic analysis; 38 of them representing other *Grifola* Spp., i.e., all available sequences of *G. colensoi* (Berk.) G. Cunn., *G. gargal* Singer, and *G. sordulenta* (Mont.) Singer, and the rest corresponding to two other species used as outgroups, i.e., *Polyporus umbellatus* (Pers.) Fr. (syn. *Grifola umbellata* (Pers.) Pilát) and *Rhodonia placenta* (Fr.) Niemelä, K.H. Larss. & Schigel (syn. *Oligoporus placentus* (Fr.) Gilb. & Ryvarden). Sequence alignment was performed through the aid of the Clustal Omega software (https://www.ebi.ac.uk/Tools/msa/clustalo/); alignments were inspected and manually adjusted at misaligned sites using MEGA X (Pennsylvania State University, State College, PA, USA) [9].

Phylogenetic relationships were inferred by using the maximum likelihood (ML) method based on the Kimura 2-parameter model [10]. Initial tree(s) for the heuristic search were obtained automatically by applying the Neighbor-Join and BioNJ algorithms to a matrix of pairwise distances estimated using the maximum composite likelihood (MCL) approach, and then selecting the topology with the superior log likelihood value. Phylogenetic analysis was conducted in MEGA X [9].

2.5. Evaluation of Nutritional Value and Determination of Antimicrobial Activity

The powder resulting after freeze-drying part of the collected basidioma was also subjected to analysis for the evaluation of the nutritional value (proximate composition, and content in elements and vitamins) according to standard methodologies reported by Palazzolo et al. [11], AOAC [12], Thompson et al. [13], and Loewus [14].

With regard to the antibacterial properties, the study initially included the evaluation of the activity of the lyophilized powder deriving from the basidioma against four Gram-positive and Gram-negative bacteria strains: *Staphylococcus aureus* ATCC 25923, *S. epidermidis* ATCC 12228, *Pseudomonas aeruginosa* ATCC 15442, and *Escherichia coli* ATCC 25922. Tryptic soy broth (Sigma-Aldrich, Darmstadt, Germany) containing glucose (2% *w/v*) or Mueller Hinton (Sigma-Aldrich, Darmstadt, Germany) was used as growth media. The lyophilized sample (3 g) was placed in a beaker with 200 mL of demineralized water, and then stored at −20 °C.

The minimum inhibitory concentration (MIC) was determined in a Minisart syringe with a 2.5-μm cellulose acetate filter by preparing solutions in vitro at increasing concentrations, incubating them with separate lots of cultivated bacteria, and measuring the results using agar dilutions. In addition, the effect of extract obtained from the basidioma powder was examined in respect to the biofilm produced by strains of *S. aureus* ATCC 43300, *Enterococcus faecalis* ATCC 29212, *P. aeruginosa* ATCC 27853, *E. coli* ATCC 25922, and *Klebsiella pneumoniae* ATCC 700603.

The bacterial strains were inoculated in Trypticase™ Soy Agar (BBL™; Becton, Dickinson and Company, Sparks, MD, USA) media and incubated at 37 °C for 24 h. In total, 100 μL of overnight grown culture (0.5 McFarland in tryptose broth, BT) was added in a sterile 96-well flat bottom microtiter plate (Biosigma S.r.l. Dominique Dutscher Group, Brumath, France). The negative control was composed by BT medium only; to demonstrate biofilm production, excess bacterial suspension was removed. Subsequently, the extract from the basidioma powder (25% *v/v*) was re-suspended in BT, 100 μL were added in the wells, and the cultures were incubated at 37 °C for 24 h. Finally, the biofilm was stained with crystal violet dissolved in ethanol (0.5% *w/v*). The optical density was measured at 540 nm (Spectrophotometer Multiskan Go; Thermo Fisher Scientific, Waltham, MA, USA). The experiment was conducted in triplicate [15,16].

3. Results and Discussion

3.1. Morphological Description of the G. frondosa Specimen and Cultivation Tests

Basidioma is composed of multiple pilei in the form of a rosette, sharing a branched stem-like structure (Figure 1a). Individual pilei were found to be more or less fan-shaped or deltoid, entirely white and yellowing with age, and finely velvety or bald, with wavy margins, 4 to 10 cm wide and 5 to 10 mm thick. The pore surface was shown to be white, staining yellowish when ripe, 2 to 3 mm deep, and irregular, and varying in shape from round to elongated (Figure 1b); it is slightly decurrent to the stipe. The stipe structure was found to be branched, whitish, tough, and often off-center. The flesh was firm, white, and unchanging when sliced. The odor and taste were found to be mild and pleasant, although less so when decaying or acrid. The chemical reaction of flesh to iodine-potassium iodide and potassium hydroxide was negative. The iron salts reaction on the pileus and flesh was negative. Basidiospores, slightly reduced when compared to the brown maitake (5–7 × 3.2 μm), were found to be broadly ellipsoidal, smooth, 5.0–7.0 × 3.2–4.0 μm, slightly narrower than those of the brown maitake, and inamyloid. Basidia were 25–30 × 6–8 μm; clavate; and 4-sterigmata (Figure 2a). Clamp connections were found to be present on generative hyphae but are rare or absent on skeletal hyphae. Hymenial cystidia and cystidioles were absent. The hyphal system is dimitic, and the spore print white. Sterigmata were found to be up to 6 μm long, and robust, with a base diameter up to 1.6 μm (Table 1). Relevant information referring to the description of common brown form [7,17] is included in Table 1. It should be mentioned that the length of the sterigmata and the diameter of their base were not reported by Ryvarden [17], while Bernicchia [7] stated that sterigmata appear "thin and very divaricate".

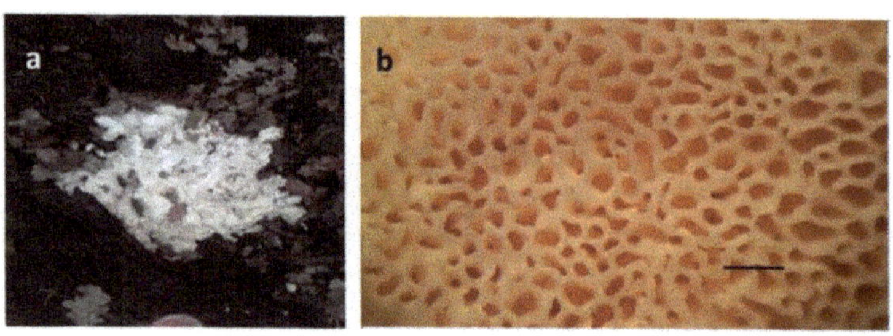

Figure 1. The albino maitake specimen: (**a**) Basidioma in situ; (**b**) Pore surface under the stereomicroscope (scale bar = 1 mm).

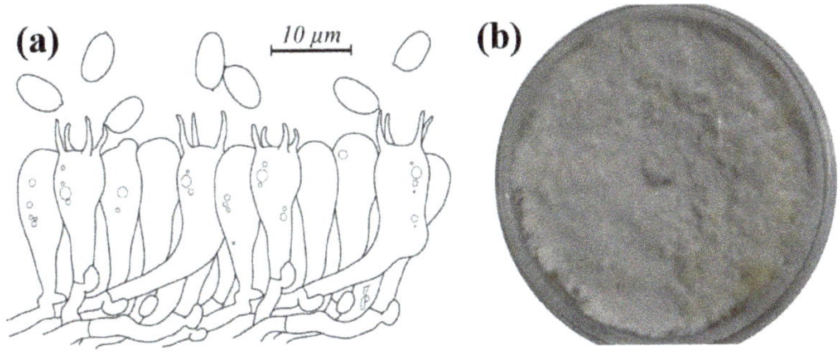

Figure 2. (**a**) Microscopic features (basidia and spores) of albino maitake. (**b**) Pure culture isolated from fresh basidioma.

Table 1. Main macro- and micromorphological features of brown [7,17] and albino (this study) forms of maitake.

Descriptive Characters	Brown Form	Albino Form
Habit	Cluster of pilei	Cluster of pilei
Pileus	Irregular, wrinkled, flat and uneven	Fan-shaped or deltoid
Stipe	Whitish	Whitish
Tubules	Short, thick, whitish	Short, thick, whitish
Pores	Very small, then angular and wider	Round to elongated, slightly decurrent to the stipe
Stem	Whitish and fleshy, with numerous flattened branches and bifurcations that end in a fan-shaped pileus	Structure branched, whitish, tough, often off-center
Flesh	White, immutable, tenacious especially towards the base, fragile towards the pileus	Firm, white, unchanging when sliced
Smell	Penetrating, intense then nauseating	Mild and pleasant, unpleasant or acrid when decaying
Taste	Pleasant and delicate	Mild and pleasant, unpleasant or acrid when decaying
Basidiospores	Ellipsoidal, 5.0–6.5 × 3.5–5.0 µm	Broadly ellipsoidal, smooth, 5.0–7.0 × 3.2–4.0 µm
Basidia	23–30 × 5–8 µm; clavate; 4-sterigmata	25–30 × 6–8 µm; clavate; 4-sterigmata
Hyphal system	Dimitic with clamp connections	Dimitic with clamp connections
Habitat	At the base of hardwoods, particularly chestnut and oak trees	At the base of oak trees
Period of fructification	Late August to October	December to January

Mycelium growth was very slow in the pure cultures established, and the surface of the Petri dish (diameter 90 mm) was completely overgrown in 30 days. The colony was found to be slightly floccose and flat, with dense white hyphae without any zonation and with a filiform margin (Figure 2b). Regarding the cultivation tests performed either in holm oak or in chestnut-based substrates, colonization was slow and ca. two months elapsed before they were completely overgrown by mycelium. At this time, a few primordia were formed on the substrate surface, which, however, failed to produce fully developed basidiomata.

3.2. *Phylogenetic Analysis*

The main objective of the molecular analysis was to identify the biological material under examination (albino maitake). Following BLAST, the sequence (MN944407) from our specimen showed high identity (>99%) to several other sequences deposited as *G. frondosa* in GenBank. The generated ML trees had similar topologies, and only the tree with the highest log likelihood (−7662.14) was selected to be presented (Figure 3).

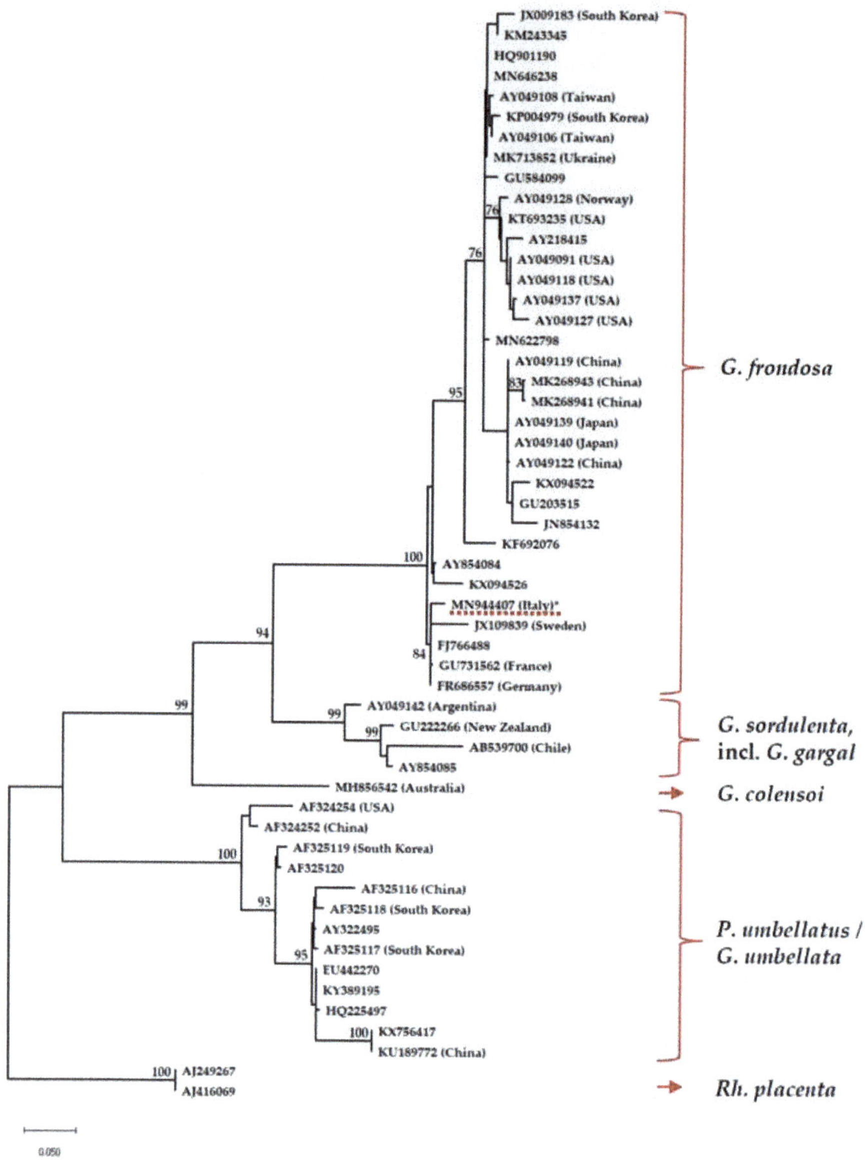

Figure 3. Maximum likelihood tree produced on the basis of ITS (internal transcribed spacer) sequence analysis presenting the phylogenetic position of the albino *G. frondosa* specimen (underlined in red and marked with an asterisk) among members of the genus *Grifola*; GenBank accession numbers are quoted together with information of their geographic origin (when available). Sequences of *Polyporus umbellatus* (syn. *G. umbellata*) and *Rhodonia placenta* were used as outgroups. The percentage of trees in which the associated taxa clustered together is shown next to the branches when ≥70% (ML bootstrap support values derived from a total of 1000 replicates). The tree is drawn to scale, with branch lengths measured as the number of substitutions per site.

To the best of our knowledge, the sequence obtained in this study is the only one available from an albino *Grifola* specimen. The outcome of the phylogenetic analysis revealed that the albino maitake clusters within the *G. frondosa* terminal clade, which receives high bootstrap support (100%), and forms a subgroup (84%) together with other sequences from material collected in Europe; the rest of the *G. frondosa* sequences derive almost exclusively from east Asia (China, Japan, South Korea, Taiwan) and the USA. Although this subgroup could represent a distinct taxonomic entity, the information generated from our work (through the examination of one basidioma and the use of a single marker only) does not allow such inferences, at least before additional specimens are included and analyzed in the frame of a multigene approach. However, it is noteworthy that *G. frondosa* sequences from Asian and eastern North American isolates were separated through the use of a partial beta-tubulin gene and ITS [18]. *G. frondosa* forms a sister group to *G. sordulenta* (94%). The latter is highly supported (99%) and consists of three sequences under this name and one representing *G. gargal* (AB539700). Hence, the distinct taxonomic position of the latter is dubious by examining the ITS marker alone. Of interest was also the position of the sole *G. colensoi* sequence available (MH856542; this species is known so far from Australia and New Zealand only), which appears to be quite distant from the rest of the material placed in the genus *Grifola*. Lastly, sequences deposited either as *G. umbellata* or *P. umbelatus* form a distinct highly supported group (100%), which is well separated from the clade corresponding to *Grifola* Spp. To the best of our knowledge, the sequence generated in the frame of this work is the first one deriving from an albino *G. frondosa* specimen.

3.3. Mushroom Proximate Composition and Content in Elements and Vitamins

In the frame of this work, it was not possible to analyze (and hence directly compare to our albino specimen) the composition of wild basidiomata of brown maitake since they were not available during the period of the study, this species being very rare in Sicily. On the other hand, only limited literature data are available referring to the chemical composition and vitamin contents in cultivated *G. frondosa* [19,20]; the respective values reported—apart from a few exceptions—are considerably lower than those found in the white maitake. However, such information should be evaluated with care since it is known that the nature of the cultivation substrate could considerably affect the mushroom content [21,22]. Similarly, metal accumulation in wild edible mushrooms was found to be influenced by various factors, including the species, the available concentration of elements in soil substrates, the substrates' properties (e.g., pH, organic matter content), the antagonistic and/or synergistic effects among elements, and the concentration of other elements in mushrooms [23–26].

The results of the proximate composition analysis of the albino maitake showed rather low values of protein, fat, and carbohydrate contents (Table 2), while the calorie content was 35.84 kcal·100 g^{-1}. In addition, element analysis showed that this specimen presents relatively high contents in Ca, Fe, K, and Cu, and is rather low in Na (Table 2). The Ca content is higher than in some baby foods, such as oatmeal and whole milk, as well as ham, cream, spread cheese, white chocolate, and others [27]. The content in Fe is considerably higher than in most other foods; it is noteworthy that dried thyme (i.e., one of the richest foods in Fe) contains about 50 mg less Fe per 100 g when compared to the albino *G. frondosa*. The K content in albino maitake is lower only to that of cremor tartar and much higher than that of dried spirulina algae [27], while the Na content is lower than that of miso soup and baking powder [27]. The Cu content of albino maitake is higher than most foods, including dark chocolate, oysters, liver, and lobster [27].

Table 2. Outcome of proximate composition analysis (% for nitrogen, proteins, fats, carbohydrates, and ash), and content in elements and vitamins (mg·100 g^{-1}) of the albino maitake specimen. Values are expressed as means ± standard error, n = 3.

Composition/Content	Albino Maitake
Nitrogen	2.18 ± 0.01
Proteins	13.65 ± 0.00
Fats	1.02 ± 0.04
Carbohydrates	6.78 ± 0.11
Ash	1.06 ± 0.10
Ca	245.30 ± 0.17
Fe	178.40 ± 0.15
Mg	455.90 ± 0.19
K	11,785.60 ± 0.36
Na	3433.40 ± 0.34
P_2O_5	2399.61 ± 0.02
Cu	2.81 ± 0.02
Mn	2.41 ± 0.03
Zn	10.33 ± 0.02
Se	4.00 ± 0.00
Pb	2.10 ± 0.00
Thiamine (B_1)	0.15 ± 0.03
Riboflavin (B_2)	3.89 ± 0.06
Niacin (B_3)	0.36 ± 0.03
Pantothenic Acid (B_5)	0.68 ± 0.02
Folic acid (B_9)	0.38 ± 0.01
D_2	0.41 ± 0.02

The content in vitamins was: B_1, 0.15 mg 100 g^{-1}; B_2, 0.36 mg·100 g^{-1}; B_3, 3.89 mg·100 g^{-1}; B_5, 0.68 mg 100 g^{-1}; B_9, 0.38 mg 100 g^{-1}; and D_2, 0.41 mg 100 g^{-1} (Table 2). Noteworthy is the value of vitamin B_9, or folic acid, since the daily requirement is about 0.2 mg, which doubles for pregnant women [28].

3.4. Antibacterial and Antibiofilm Activity

The minimal inhibitory concentration (MIC) of the extract deriving from the albino maitake was determined by a micro-method, i.e., by diluting the protein extract in the range from 25% to 0.025% *v/v* as previously described [29].

The antibacterial activity test carried out on albino maitake showed that the cold water extract was effective in inhibiting the growth of two bacterial strains, i.e., *S. epidermidis* ATCC 12228 and *P. aeruginosa* ATCC 15442, at the maximum screening concentration of 50% *v/v*. *S. epidermidis* is a Gram-positive bacterium typically present on the skin, which could be particularly dangerous in surgical procedures, while *P. aeruginosa*, is a ubiquitous bacterium and opportunistic pathogen in humans. As regards the effect of the water extract from the albino *G. frondosa* on biofilm formation, the results revealed that it decreased the biofilm produced by *S. aureus* ATCC43300 [30], whereas it slightly enhanced biofilm formation by *E. faecalis* ATCC 29212, *E. coli* ATCC 25922, *K. pneumoniae* ATCC 700603, and *P. aeruginosa* ATCC 15442 (Figure 4).

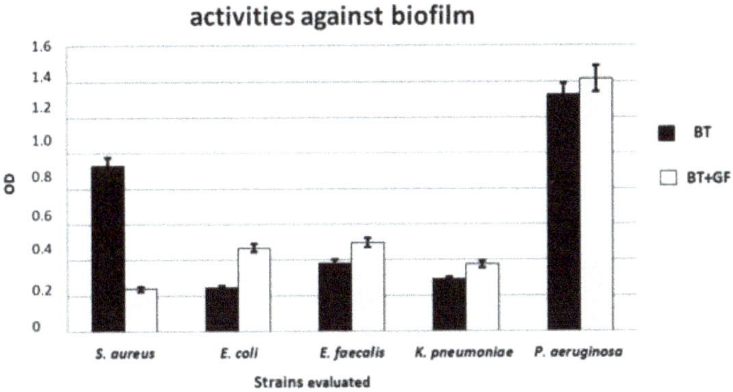

Figure 4. Effect of the cold water extract from the albino *Grifola frondosa* basidioma on biofilm production (expressed as optical density, OD) by five bacterial species. BT: biofilm produced in the absence of the fungal extract; BT+GF: biofilm produced in the presence of the fungal extract. The experiment was conducted in triplicate.

3.5. Ecological and Conservation Issues

The characterization of mushrooms growing in the Mediterranean area is one of the main targets of studies carried out by our research group. The principal objective is to provide reliable/robust scientific data amid the considerable confusion existing in the commercial exploitation of pertinent material in Italy and elsewhere, since accurate identification and/or origin of products placed on the market are often missing or of ambiguous validity. Previous results have highlighted the importance of mushrooms of the genus *Pleurotus* in human nutrition and in the feeding of farm animals [31–35]. Particular attention is also drawn to edible mushroom species that are infrequent/rare or linked to particular habitats (e.g., *Pleurotus nebrodensis* and *P. opuntiae*) [36,37], and for which only few or no studies exist regarding their nutritional and medicinal properties.

Although *G. frondosa* is widely distributed in the northern hemisphere, it is considered very rare in Italy. Particularly in Sicily, the only reports available concern the brown maitake found in the northeastern part of the island and date back to the end of the 19th century [38]; more recent informal records by amateur mycological groups derive from the same wider region (however, no pertinent samples are available for further study). Hence, the occurrence of an albino maitake is of apparent importance in a biogeographical, ecological, and applied context. In addition, the presence of this fungus on a monumental *Q. pubescens* tree of high significance points out the potential risks for host plants in the area because *G. frondosa* demonstrates a parasitic action resulting in root and butt decay [39,40]. At the same time, it is necessary to ensure the necessary conditions for its ex situ conservation due to the exploitation potential it exhibits as food and source of bioactive compounds as it was reported for wild mushrooms in general [41]. We have therefore adopted well-established protocols for its long-term maintenance [42] since our investigation showed that fructification of the albino maitake is not constant over time and the ability of the strain to be preserved under laboratory conditions—without significant degeneration of its properties—is lower than in other mushroom species.

4. Conclusions

An albino maitake specimen was reported for the first time in Italy. The examined basidioma differs morphologically from the common brown/gray form, with the most significant differences being the color and shape of the pileus, the type of pores, the shape of the stipe, and the period of fructification. Phylogenetic analysis evidenced that the specimen under study groups among other European isolates within a larger terminal clade corresponding to *G. frondosa*. Moreover, the nutritional composition and the potential of the albino maitake to serve as source of functional compounds is of interest, and further investigation is in progress on the characterization of polysaccharides and at evaluating the antitumor activity against several types of cancer cells. The outcome of tests performed using the mushroom water extract against human pathogenic bacteria demonstrated its efficacy at inhibiting the growth of *S. epidermidis* and *P. aeruginosa* strains, while it also hindered biofilm formation by *S. aureus*, i.e., the causal agent of serious infections in immunocompromised patients. The antibacterial agents present in the albino maitake are promising candidates for dealing with pathogens that show resistance to a large spectrum of antibiotics and are widespread in hospitals.

Author Contributions: Conceptualization, G.V. and M.L.G.; methodology, G.V., G.I.Z., M.L.G. and O.S.I.; validation, R.C. and V.F.; formal analysis, G.I.Z., M.L.G. and O.S.I.; investigation, G.V., G.I.Z., M.L.G. and O.S.I.; data curation, A.G. and T.F.; biofilm analysis, R.C. and V.F.; writing—original draft preparation, G.V. and M.L.G.; writing—review and editing, G.V. and G.I.Z.; visualization, G.V.; supervision, G.V. All authors have read and agreed to the published version of the manuscript.

Funding: This research received no external funding.

Conflicts of Interest: The authors declare no conflicts of interest.

References

1. Justo, A.; Miettinen, O.; Floudas, D.; Ortiz-Santana, B.; Sjökvist, E.; Lindner, D.L.; Nakasone, K.; Niemelä, T.; Larsson, K.-H.; Ryvarden, L.; et al. A revised family-level classification of the Polyporales (*Basidiomycota*). *Fungal Biol.* **2017**, *121*, 798–824. [CrossRef] [PubMed]
2. Chen, A.W.; Stamets, P.; Cooper, R.B.; Huang, N.L.; Han, S.-H. Ecology, morphology, and morphogenesis in nature of edible and medicinal mushroom *Grifola frondosa* (Dicks.: Fr.) S.F. Gray—Maitake (*Aphyllophoromycetideae*). *Int. J. Med. Mushrooms* **2000**, 2, 8. [CrossRef]
3. Imazeki, R. Genera of *Polyporaceae* of Nippon. *Bull. Nat. Sci. Mus. Tokyo* **1943**, *6*, 1–111.
4. Kawaguchi, N.; Hayashi, M.; Chen, F.-C.; Shimomura, N.; Yamaguchi, T.; Aimi, T. Genetic analyses of causal genes of albinism (white fruiting body) in *Grifola frondosa*. *J. Wood Sci.* **2019**, *65*, 32. [CrossRef]
5. Kawaguchi, N.; Hayashi, M.; Nakano, S.; Shimomura, N.; Yamaguchi, T.; Aimi, T. Expression of tyrosinase genes associated with fruiting body formation and pigmentation in *Grifola frondosa*. *Mycoscience* **2019**, *60*, 262–269. [CrossRef]
6. Gargano, M.L.; Van Griensven, L.J.L.D.; Isikhuemhen, O.S.; Lindequist, U.; Venturella, G.; Wasser, S.P.; Zervakis, G.I. Medicinal mushrooms: Valuable biological resources of high exploitation potential. *Plant Biosyst.* **2017**, *151*, 548–565. [CrossRef]
7. Bernicchia, A. *Polyporaceae s.l.*; Edizioni Candusso: Alassio, Italy, 2005; p. 808.
8. White, T.J.; Bruns, T.; Lee, S.; Taylor, J. *Amplification and Direct Sequencing of Fungal Ribosomal RNA Genes for Phylogenetics*; Academic Press: Cambridge, MA, USA, 1990; pp. 315–322.
9. Kumar, S.; Stecher, G.; Li, M.; Knyaz, C.; Tamura, K. MEGA X: Molecular evolutionary genetics analysis across computing platforms. *Mol. Biol. Evol.* **2018**, *35*, 1547–1549. [CrossRef]
10. Kimura, M. A simple method for estimating evolutionary rates of base substitutions through comparative studies of nucleotide sequences. *J. Mol. Evol.* **1980**, *16*, 111–120. [CrossRef]
11. Palazzolo, E.; Gargano, M.L.; Venturella, G. The nutritional composition of selected wild edible mushrooms from Sicily (southern Italy). *Int. J. Food Sci. Nutr.* **2011**, *63*, 79–83. [CrossRef]
12. AOAC. Official Methods of Analysis of the AOAC. *Anal. Chem.* **1980**, *52*, 148A. [CrossRef]
13. Thompson, T.E.; Senter, S.D.; Grauke, L. Lipid content and fatty acids of pecan pollen. *HortScience* **1993**, *28*, 1191–1193. [CrossRef]
14. Loewus, F.A. Improvement in anthrone method for determination of carbohydrates. *Anal. Chem.* **1952**, *24*, 219. [CrossRef]
15. Calà, C.; Amodio, E.; Di Carlo, E.; Virruso, R.; Fasciana, T.; Giammanco, A. Biofilm production in *Staphylococcus* epidermidis strains, isolated from the skin of hospitalized patients: Genetic and phenotypic characteristics. *New Microbiol.* **2015**, *38*, 521–529. [PubMed]

16. Fasciana, T.; Gentile, B.; Aquilina, M.; Ciammaruconi, A.; Mascarella, C.; Anselmo, A.; Fortunato, A.; Fillo, S.; Petralito, G.; Lista, F.; et al. Co-existence of virulence factors and antibiotic resistance in new *Klebsiella pneumoniae* clones emerging in south of Italy. *BMC Infect. Dis.* **2019**, *19*, 928. [CrossRef]
17. Ryvarden, L.; Gilbertson, R.L. European Polypores. Part 1. *Synopsis Fungorum* **1993**, *6*, 1–387.
18. Shen, Q.; Geiser, D.M.; Royse, D.J. Molecular Phylogenetic Analysis of *Grifola frondosa* (Maitake) reveals a species partition separating Eastern North American and Asian Isolates. *Mycologia* **2002**, *94*, 472–482. [CrossRef]
19. Stamets, P. Notes on Nutritional properties of culinary-medicinal mushrooms. *Int. J. Med. Mushrooms* **2005**, *7*, 103–110. [CrossRef]
20. Niedzielski, P.; Mleczek, M.; Budka, A.; Rzymski, P.; Siwulski, M.; Jasińska, A.; Gąsecka, M.; Budzyńska, S. A screening study of elemental composition in 12 marketable mushroom species accessible in Poland. *Eur. Food Res. Technol.* **2017**, *243*, 1759–1771. [CrossRef]
21. Koutrotsios, G.; Mountzouris, K.; Chatzipavlidis, I.; Zervakis, G.I. Bioconversion of lignocellulosic residues by *Agrocybe cylindracea* and *Pleurotus ostreatus* mushroom fungi—Assessment of their effect on the final product and spent substrate properties. *Food Chem.* **2014**, *161*, 127–135. [CrossRef]
22. Sakellari, A.; Karavoltsos, S.; Tagkouli, D.; Rizou, C.; Sinanoglou, V.J.; Zoumpoulakis, P.; Koutrotsios, G.; Zervakis, G.I.; Kalogeropoulos, N. Trace elements in *Pleurotus ostreatus*, *P. eryngii*, and *P. nebrodensis* Mushrooms Cultivated on various agricultural by-products. *Anal. Lett.* **2019**, *52*, 2692–2709. [CrossRef]
23. Garcia, E.M.; Siegert, I.G.; Suarez, P. Toxicity assays and naphthalene utilization by natural bacteria selected in marine environments. *Bull. Environ. Contam. Toxicol.* **1998**, *61*, 370–377. [CrossRef] [PubMed]
24. Mendil, D.; Özgür, D.U.; Tuzen, M.; Hasdemir, E.; Sarı, H.; Sari, H. Trace metal levels in mushroom samples from Ordu, Turkey. *Food Chem.* **2005**, *91*, 463–467. [CrossRef]
25. Venturella, G.; Gargano, M.L.; Compagno, R.; Saitta, A.; Alaimo, M.G. The mineral contents of some boletaceae species from Sicily (Southern Italy). *J. AOAC Int.* **2014**, *97*, 612–623. [CrossRef]
26. Kokkoris, V.; Massas, I.; Polemis, E.; Koutrotsios, G.; Zervakis, G.I. Accumulation of heavy metals by wild edible mushrooms with respect to soil substrates in the Athens metropolitan area (Greece). *Sci. Total. Environ.* **2019**, *685*, 280–296. [CrossRef] [PubMed]
27. Fidanza, F. Alimenti e Tabelle di Composizione. In *Nutrizione Umana*; Fidanza, F., Liguori, G., Eds.; Idelson: Napoli, Italy, 1984; pp. 221–238.
28. Barua, S.; Kuizon, S.; Junaid, M.A. Folic acid supplementation in pregnancy and implications in health and disease. *J. Biomed. Sci.* **2014**, *21*, 77. [CrossRef] [PubMed]
29. Schillaci, D.; Petruso, S.; Sciortino, V. 3,4,5,3′,5′-Pentabromo-2-(2′-hydroxybenzoyl) pyrrole: A potential lead compound as anti-Gram-positive and anti-biofilm agent. *Int. J. Antimicrob. Agents* **2005**, *25*, 338–340. [CrossRef] [PubMed]
30. Giammanco, A.; Galia, E.; Fasciana, T.; Sciortino, M.; Cannella, S.; Arrigo, I.; Tricoli, M.R.; Di Paola, L.; Venturella, G.; Gargano, M.L. The Anti-Biofilm Potential of Pleurotus eryngii var. elaeoselini and White form of Grifola frondosa Extracts Against Staphylococcus aureus methicillin Resistance. In Proceedings of the 10th International Medicinal Mushrooms Conference, Nantong, China, 19–22 September 2019; International Society for Medicinal Mushrooms & Jiangsu Provincial Department of Agriculture and Rural Affairs: Nantong, China, 2019; pp. 184–185.
31. La Guardia, M.; Venturella, G.; Venturella, F. On the chemical composition and nutritional value of *Pleurotus taxa* growing on Umbelliferous plants (*Apiaceae*). *J. Agric. Food Chem.* **2005**, *53*, 5997–6002. [CrossRef] [PubMed]
32. Koutrotsios, G.; Kalogeropoulos, N.; Stathopoulos, P.; Kaliora, A.C.; Zervakis, G.I. Bioactive compounds and antioxidant activity exhibit high intraspecific variability in *Pleurotus ostreatus* mushrooms and correlate well with cultivation performance parameters. *World J. Microbiol. Biotechnol.* **2017**, *33*, 98. [CrossRef]
33. Koutrotsios, G.; Kalogeropoulos, N.; Kaliora, A.C.; Zervakis, G.I. Toward an Increased functionality in oyster (*Pleurotus*) mushrooms produced on Grape Marc or Olive Mill Wastes Serving as sources of bioactive compounds. *J. Agric. Food Chem.* **2018**, *66*, 5971–5983. [CrossRef]
34. Gaglio, R.; Guarcello, R.; Venturella, G.; Palazzolo, E.; Francesca, N.; Moschetti, G.; Settanni, L.; Saporita, P.; Gargano, M.L. Microbiological, chemical and sensory aspects of bread supplemented with different percentages of the culinary mushroom *Pleurotus eryngii* in powder form. *Int. J. Food Sci. Technol.* **2018**, *54*, 1197–1205. [CrossRef]

35. Bonanno, A.; Di Grigoli, A.; Vitale, F.; Di Miceli, G.; Todaro, M.; Alabiso, M.; Gargano, M.L.; Venturella, G.; Anike, F.N.; Isikhuemhen, O.S. Effects of diets supplemented with medicinal mushroom myceliated grains on some production, health, and oxidation traits of dairy ewes. *Int. J. Med. Mushrooms* **2019**, *21*, 89–103. [CrossRef]
36. Zervakis, G.I.; Ntougias, S.; Gargano, M.L.; Besi, M.I.; Polemis, E.; Typas, M.A.; Venturella, G. A reappraisal of the *Pleurotus eryngii* complex—New species and taxonomic combinations based on the application of a polyphasic approach, and an identification key to *Pleurotus taxa* associated with *Apiaceae plants*. *Fungal Biol.* **2014**, *118*, 814–834. [CrossRef] [PubMed]
37. Zervakis, G.I.; Venturella, G.; Fryssouli, V.; Inglese, P.; Polemis, E.; Gargano, M.L. *Pleurotus opuntiae* revisited—An insight to the phylogeny of dimitic *Pleurotus* species with emphasis on the P. djamor complex. *Fungal Biol.* **2019**, *123*, 188–199. [CrossRef] [PubMed]
38. Venturella, G. A check-list of Sicilian fungi. *Bocconea* **1991**, *2*, 1–221.
39. Antonio, A.; Valeria, B.; Marcello, D.; Caterina Patrizia Di, F.; Di Gesaro, M.; Emanuele, S. Monumental trees and their existence value: The case study of an Italian natural park. *J. For. Sci.* **2016**, *61*, 55–61. [CrossRef]
40. Schwarze, F.W.M.R.; Engels, J.; Mattheck, C. *Fungal Strategies of Wood Decay in Trees*; Springer: Berlin/Heidelberg, Germany, 2000.
41. Zotti, M.; Persiani, A.M.; Ambrosio, E.; Vizzini, A.; Venturella, G.; Donnini, D.; Angelini, P.; Di Piazza, S.; Pavarino, M.; Lunghini, D.; et al. Macrofungi as ecosystem resources: Conservation versus exploitation. *Plant Biosyst.* **2013**, *147*, 219–225. [CrossRef]
42. Moore, D.; Nauta, M.M.; Evans, S.E.; Rotheroe, M. *Fungal Conservation. Issues and Solutions*; Cambridge University Press: Cambridge, UK, 2001; p. 262.

diversity

MDPI

Article

An Italian Research Culture Collection of Wood Decay Fungi

Carolina Elena Girometta [1,*], Annarosa Bernicchia [2], Rebecca Michela Baiguera [1], Francesco Bracco [1], Simone Buratti [1], Marco Cartabia [1,3], Anna Maria Picco [1] and Elena Savino [1]

1 Department of Earth and Environmental Sciences (DSTA), University of Pavia, 27100 Pavia, Italy; rebeccamichela.baiguera01@universitadipavia.it (R.M.B.); francesco.bracco@unipv.it (F.B.); simone.buratti01@universitadipavia.it (S.B.); marco.cartabia01@universitadipavia.it (M.C.); annamaria.picco@unipv.it (A.M.P.); elena.savino@unipv.it (E.S.)

2 School of Agriculture and Veterinary Medicine, University of Bolognavia, Guidotti 39, 40134 Bologna, Italy; corticia.polypores@gmail.com

3 MOGU S.r.l. Via San Francesco d'Assisi 62, 21020 Inarzo (VA), Italy

* Correspondence: carolinaelena.girometta@unipv.it

Received: 29 December 2019; Accepted: 30 January 2020; Published: 1 February 2020

Abstract: One of the main aims of the University of Pavia mycology laboratory was to collect wood decay fungal (WDF) strains in order to deepen taxonomic studies, species distribution, officinal properties or to investigate potential applications such as biocomposite material production based on fungi. The Italian Alps, Apennines and wood plains were investigated to collect Basidiomycota basidiomata from living or dead trees. The purpose of this study was to investigate the wood decay strains of the Mediterranean area, selecting sampling sites in North and Central Italy, including forests near the Ligurian and Adriatic seas, or near the Lombardy lakes. The isolation of mycelia in pure culture was performed according to the current methodology and the identity of the strains was confirmed by molecular analyses. The strains are maintained in the Research Culture Collection MicUNIPV of Pavia University (Italy). Among the 500 WDF strains in the collection, the most interesting isolates from the Mediterranean area are: *Dichomitus squalens* (basidioma collected from *Pinus pinea*), *Hericium erinaceus* (medicinal mushroom), *Inocutis tamaricis* (white-rot agent on *Tamarix* trees), *Perenniporia meridionalis* (wood degrader through Mn peroxidase) and *P. ochroleuca*. In addition, strains of species related to the Mediterranean climate (e.g., *Fomitiporia mediterranea* and *Cellulariella warnieri*) were obtained from sites with a continental-temperate climate.

Keywords: wood decay fungi (WDF); culture collection; fungal strain; host; Italy; morphological and molecular identification

1. Introduction

Wood decay fungi provide an extraordinary model both for pure and applied research, as well as a food or medicinal mushroom resource.

From an ecological point of view, wood decay fungi have a fundamental role, since they are important degraders of lignocelluloses. Heterogeneity in degradation strategies consists in different enzymatic pools, conditions for secretion and catalysis, alternative non-enzymatic pathways and strategy-switch depending on environmental conditions [1,2]. Consistently, wood decay fungi shift from necrotrophism to pure saprotrophism, sometimes at an intraspecific level.

Systematic revisions based on multi-locus or genomic approach have revealed an extremely complex scenario concerning both biochemical features and morphology. Similar degradation strategies and similar morphologies are widespread, even among phylogenetically distant taxa, whereas the same taxonomic group may include species displaying different strategies and different morphologies.

As a whole, wood decayers appear to be a pivotal model in the study of the evolutionary relationships of both Dikarya and extra-Dikarya taxa [3–5]. Consistently, the incipient molecular-based biogeography of wood decayers seems to display distribution patterns strongly affected by preferred host species (trees or shrubs), which are apparently followed throughout [6].

Wood decay fungi include several edible species as well as species reported to be the source of bioactive compounds, related to either primary (e.g., β-glucans) or secondary metabolism (terpenoids, phenolics, acids, superior alcohols, etc.). Structural diversity, occurrence and distribution among taxa, synthesis stimulation factors and correlation to growth stage, bioactivity pathways, standardization of products and crude extracts are the main current topics under investigation [7–9].

Due to their relatively easy reproduction in culture, several other applications are being developed on wood decay fungi: degradation of organic pollutants and bioremediation [10]; pretreatment of biomasses for production of sugars and bioethanol [11,12]; production of enzymes for industrial purposes, namely Mn peroxidases, laccases, cellulases and hemicellulases [13]; and bioadsorption and bioaccumulation of metal ions either in living or dead biomass [14–16].

Last but not least, necrotrophic wood decay means a loss of harvest in forestry and woody cultures, whereas in public and private green areas it means destabilization and consequently risk for people and objects [17].

Culture collections are an important reference, since availability and exchange of authenticated, quality-guaranteed pure cultures are increasingly needed by researchers at an international level [18,19]. Above all, tests on different species and strains (at intraspecific level) are required since biochemical differences are often not negligible [20–22].

Actually, only a few research centers can afford to structure their culture collection in conformity to international guidelines provided by the World Federation for Culture Collections (WFCC) [23,24]. The strains maintained in many universities or research centers can be considered an important source of experimental material, even without a WFCC certification. Since small uncertified collectors are geographically widespread, their contribution may be significantly representative of local ecosystems and biodiversity [25].

The Mycology Laboratory at the Botanical Garden of Pavia University (Italy) has a long tradition of the isolation, identification and preservation of fungal strains in various areas of mycology. This is supported by the numerous publications from the middle of the last century [26] up to now. Currently, the fungal strains collection is named MicUNIPV and each working group preserves and enriches the collection.

Although the definition of ecotypes is usually hard, a remarkable intraspecific variability is well documented in several fungal species and it may be particularly true for rare species, whose populations are supposed to be more isolated [27]. This highlights the value of Italian territory for fungal biodiversity and the great potentiality for research [28].

Italy has a wide variety of climates and morphologies, both due to its remarkable latitudinal range (about 13°) and structural-topographic complexity, including the presence of four different seas and two main mountain chains. According to the official maps of MATTM (Ministero dell'Ambiente e della Tutela del Territorio e del Mare), 28 different phytoclimatic classes are recognized, five of which are specifically referred to as Mediterranean [29], also taking into account the biogeographic reference map suggested by Rivas-Martínez [30]. Nevertheless, the pluri-millennial stratification of human impact has made it difficult to distinguish between actual and potential ecosystem features. As a consequence, the classification and mapping of either Italian ecoregions or phytoclimates provide a tool for the comprehension of biodiversity instead of a strict map of biodiversity itself [31].

The present article reports the results obtained by the researchers of the Laboratory of Mycology in DSTA-University of Pavia (Italy) who continuously collect new cultures of wood decay fungi, focusing on fungal biodiversity of species related to the Mediterranean area and climates.

2. Materials and Methods

2.1. Sampling Sites and Field Work

Basidiomata were mostly (but not exclusively) collected in North and Central Italy. Sampling stratification was selectively applied, i.e., specific areas have been more frequently and strictly examined than others and sampling effort was not equal among different species [32]. The different environments examined are resumed hence:

(a) highly-fragmented marginal woodlands and shrublands placed into an agricultural landscape, particularly referring to vegetation surrounding the hydrographic network, including major lakes;
(b) mountain continuous woodlands and shrublands, both managed and unmanaged;
(c) woody cultures (e.g., poplar plantations and vineyards), tree rows and hedges in agricultural landscape;
(d) urban and suburban environments (tree rows, parks, private and public gardens).

Environments a, c, and d are mostly related to basal altitudinal belt and upper hill altitudinal belt in Po Plain, Apennines and Prealps (lower mountain thermal belt), as well as Adriatic, Tirrenian and Ligurian coasts; Environment b is mostly related to the lower and upper montane belt in the North and Central Apennines.

The basidiomata were completely or partially harvested by knife, gently brushed to eliminate debris and stored in paper bags until laboratory operations. The collecting sites were geolocalized, and the host species and general features were detected.

2.2. Experimental Procedures

Basidiomata identification was carried out by macro and micro-morphological analysis [6,33,34]; stereo and light microscopy were performed by Zeiss Axioplan and Zeiss Stemi 2000-C.

According to Stalpers [35] and Gams et al. [36], as well as Stamets [37], isolation of mycelia in pure culture was obtained in sterile conditions by inoculating small portions of the basidioma context into Petri dishes containing MEA medium and antibiotic (malt extract 2% + agar 1.5% + cloramphenicol 50 ppm). The incubation was carried out at 24 °C in the dark and each strain growth was checked constantly for a month. Based on the above, all the mentioned strains are to be regarded as dikaryotic.

Besides the morphological checks, molecular identifications of isolates were carried out on mycelia cultured in liquid medium (malt extract 2%). DNA was extracted from lyophilized mycelia by Nucleospin Plant II kit (Macherey-Nagel). Amplification by Polymerase Chain Reaction (PCR) used the primer pair ITS1 (19bp) and ITS4 (20bp)—that is, Internal Transcribed Spacer of ribosomal DNA; this region has been widely used for different fungal taxa [38,39]. PCR protocol exploited Dream Taq Mastermix (Promega) and was performed in a thermocycler, as reported in Table 1.

Table 1. Thermocycling protocol for PCR.

Step	Aim	T (°C)	Duration	Cycle Repetitions
I	Denaturation	95	5 min	
	Denaturation	95	30 s	
II	Annealing	50	45 s	35
	Elongation	72	1 min	
III	Final elongation	72	10 min	

The qualitative checking of DNA (5 µL/sample) was performed both after extraction and amplification by DNA run (30 min, 100 V) on electrophoretic gel (1% agarose). SYBR Safe-DNA Gel Stain (Invitrogen) was used as an intercalant; GeneRuler 1kb (Thermo Scientific, Waltham - USA) was used as a ladder; BlueJuice (Invitrogen) was used as a gel loading buffer. The imaging was performed by Gel Doc (Biorad, Berkeley, CA, USA).

ExoSAP-IT (Applied Biosystems, Foster City, CA, USA) was used for the purification of amplification products. According to the suggested protocol, the sample/ExoSAP ratio was 5:2 μL; the reaction was carried out in a thermocycler in two steps—15 min at 37 °C and 15 min at 80 °C.

The sequencing was ordered to Macrogen (The Netherlands). Sequence analysis was performed by Sequencher 5.0 Demo. The sequences were finally matched with the ones available in the molecular identification facility of Mycobank [40].

Strains in pure culture were stored by different methods:

(a) on malt extract agar (MEA) in a aPetri plate at 3 °C;
(b) in a glass tube corked with cotton at room temperature;
(c) colonized paper discs in demineralized water at 4 °C;
(d) at −80 °C in glycerol (selected strains only).

Periodic checking and refreshment of cultures was performed to avoid contamination and devitalization.

The strains are maintained in the Fungal Research Culture Collection (MicUNIPV) of Department of Earth and Environmental Sciences of University of Pavia (Italy); each strain is included in a private database with all the information regarding sampling sites, data of collection and ecological notes.

3. Results and Discussion

MicUNIPV includes species related to plant pathology, soil, extreme environments, fresh and marine water, monuments and cultural heritage. As previously mentioned, different working groups within the Laboratory of Mycology (DSTA-University of Pavia) are engaged in the management, preservation and improvement of each MicUNIPV section. The section regarding wood decay species has up to now achieved 500 strains belonging to 110 different species [41–43]. The broad focus on wood decay led us to include in this section species related to different applications such as nutraceuticals, forest pathology, wood degradation and biocomposite materials.

The distribution of most species exceeds the Mediterranean area; nevertheless, several of them also display wide spatial gaps among stations and clear heterogeneity in host preference depending on the geographic location of the population.

Here, we present the species that have a distribution strongly related to the Mediterranean region and/or Southern Europe and/or warm climates, according to Ryvarden and Melo [6] and Bernicchia [33,34,44]. The species related to the Mediterranean diversity are reported in Table 2 and the most peculiar are discussed below.

Table 2. Selected Italian strains from MicUNIPV related to the Mediterranean area. Phytoclimate class as in [29].

Mic UNIPV ID	Species	Authors	Locality	Municipality	Host	Phytoclimate Class
D.con.1	*Daedaleopsis confragosa*	(Bolton) J. Schröt.	Dormelletto	Dormelletto (NO)	*Unidentified broadleaf*	mesotemperate/humid supratemperate
D.con.2	*Daedaleopsis confragosa*	(Bolton) J. Schröt.	Pian Porcino	Bagno di Romagna (FC)	*Unidentified broadleaf*	hyperhumid supratemperate/ultrahyperhumid
D.q.1	*Daedalea quercina*	(L.) Pers.	R.N. Bosco Giuseppe Negri	Pavia (PV)	*Quercus robur*	humid supratemperate/subhumid
D.q.2	*Daedalea quercina*	(L.) Pers.	Cono di Volo Malpensa	Gallarate (VA)	*Quercus rubra*	mesotemperate/humid supratemperate
D.q.3	*Daedalea quercina*	(L.) Pers.	Fosso dell'Oca	Rovescala (PV)	*Quercus petraea*	humid supratemperate/subhumid
D.sq.1	*Dichomitus squalens*	(P. Karst.) D.A. Reid	Pineta di San Vitale	Ravenna (RA)	*Pinus pinea*	supratemperate/humid-subhumid mesotemperate
D.sq.2	*Dichomitus squalens*	(P. Karst.) D.A. Reid	Ispra, Lungolago	Ispra (VA)	*Cedrus sp.*	mesotemperate/humid supratemperate
D.tric.1	*Daedaleopsis tricolor*	(Bull.) Bondartsev and Singer	Rio Bardonezza	Santa Maria della Versa (PV)	*Prunus avium*	humid supratemperate/subhumid
Des.t.1	*Desarmillaria tabescens*	(Scop.) R.A. Koch and Aime	RNIS Bosco Siro Negri	Zerbolò (PV)	*Quercus robur*	humid supratemperate/subhumid
Fm.i.1	*Fomitopsis iberica*	Melo and Ryvarden	Via Montello	Varese (VA)	*Corylus avellana*	humid supratemperate/hyperhumid
Fm.i.2	*Fomitopsis iberica*	Melo and Ryvarden	Villa Baragiola	Varese (VA)	*Abies alba*	humid supratemperate/hyperhumid
Fm.i.3	*Fomitopsis iberica*	Melo and Ryvarden	Via Tasso	Varese (VA)	*Cedrus deodara*	humid supratemperate/hyperhumid
Fm.i.4	*Fomitopsis iberica*	Melo and Ryvarden	Via S. Francesco	Inarzo (VA)	*Betula pendula*	humid supratemperate/hyperhumid
Fm.i.5	*Fomitopsis iberica*	Melo and Ryvarden	Villa Toeplitz	Varese (VA)	*Fagus sylvatica*	humid supratemperate/hyperhumid
Fm.i.6	*Fomitopsis iberica*	Melo and Ryvarden	Villa Mylius	Varese (VA)	*Fagus sylvatica*	humid supratemperate/hyperhumid
Fm.m.1	*Fomitiporia mediterranea*	M. Fisch.	RNIS Bosco Siro Negri	Zerbolò (PV)	*Quercus robur*	humid supratemperate/subhumid
Fm.m.2	*Fomitiporia mediterranea*	M. Fisch.	RNIS Bosco Siro Negri	Zerbolò (PV)	*Quercus robur*	humid supratemperate/subhumid
Fm.m.3	*Fomitiporia mediterranea*	M. Fisch.	RNIS Bosco Siro Negri	Zerbolò (PV)	*Hedera helix*	humid supratemperate/subhumid
Fm.m.4	*Fomitiporia mediterranea*	M. Fisch.	Rio Bardonezza	Santa Maria della Versa (PV)	*Robinia pseudoacacia*	humid supratemperate/subhumid
Fm.m.5	*Fomitiporia mediterranea*	M. Fisch.	Comiso	Ragusa (RG)	*Cistus sp.*	mesomediterranean/subhumid-dry thermomediterranean
Fm.m.6	*Fomitiporia mediterranea*	M. Fisch.	Cono di Volo Malpensa	Gallarate (VA)	*Quercus rubra*	mesotemperate-humid supratemperate
Fm.m.7	*Fomitiporia mediterranea*	M. Fisch.	R.N. Torbiere del Sebino	Provaglio d'Iseo (BS)	*Corylus avellana*	mesotemperate-humid supratemperate
Fm.m.8	*Fomitiporia mediterranea*	M. Fisch.	Villa Augusta	Varese (VA)	*Fagus sylvatica*	humid supratemperate - hyperhumid
Fm.m.9	*Fomitiporia mediterranea*	M. Fisch.	Olgiate Comasco	Olgiate Comasco (CO)	*Actinidia chinensis*	mesotemperate-humid supratemperate

Table 2. *Cont.*

Mic UNIPV ID	Species	Authors	Locality	Municipality	Host	Phytoclimate Class
Fm.m.10	*Fomitiporia mediterranea*	M. Fisch.	Pradone nord	Rovescala (PV)	*Vitis vinifera*	humid supratemperate - subhumid
Fm.m.11	*Fomitiporia mediterranea*	M. Fisch.	Pradone nord	Rovescala (PV)	*Vitis vinifera*	humid supratemperate - subhumid
Fm.m.12	*Fomitiporia mediterranea*	M. Fisch.	Pradone nord	Rovescala (PV)	*Vitis vinifera*	humid supratemperate - subhumid
Fm.m.13	*Fomitiporia mediterranea*	M. Fisch.	Pradone nord	Rovescala (PV)	*Vitis vinifera*	humid supratemperate - subhumid
Fm.m.14	*Fomitiporia mediterranea*	M. Fisch.	Pradone nord	Rovescala (PV)	*Vitis vinifera*	humid supratemperate - subhumid
Fm.m.15	*Fomitiporia mediterranea*	M. Fisch.	Pradone nord	Rovescala (PV)	*Vitis vinifera*	humid supratemperate - subhumid
Fm.m.16	*Fomitiporia mediterranea*	M. Fisch.	Pradone nord	Rovescala (PV)	*Vitis vinifera*	humid supratemperate - subhumid
Fm.m.17	*Fomitiporia mediterranea*	M. Fisch.	Unipv_polo scientifico via Ferrata	Pavia (PV)	*Salix alba*	humid supratemperate - subhumid
G.adsp.1	*Ganoderma adspersum*	(Schulzer) Donk	Orto Botanico	Pavia (PV)	*Quercus sp.*	humid supratemperate - subhumid
G.adsp.2	*Ganoderma adspersum*	(Schulzer) Donk	R.N. Bosco Giuseppe Negri	Pavia (PV)	*Populus nigra*	humid supratemperate - subhumid
G.adsp.3	*Ganoderma adspersum*	(Schulzer) Donk	R.N. Bosco Giuseppe Negri	Pavia (PV)	*Unidentified broadleaf*	humid supratemperate - subhumid
G.adsp.4	*Ganoderma adspersum*	(Schulzer) Donk	Parco della Vernavola	Pavia (PV)	*Alnus glutinosa*	humid supratemperate - subhumid
G.adsp.5	*Ganoderma adspersum*	(Schulzer) Donk	RNIS Bosco Siro Negri	Zerbolò (PV)	*Unidentified broadleaf*	humid supratemperate - subhumid
G.adsp.6	*Ganoderma adspersum*	(Schulzer) Donk	Rio Bardonezza	Rovescala (PV)	*Quercus robur*	humid supratemperate - subhumid
G.adsp.7	*Ganoderma adspersum*	(Schulzer) Donk	Madonna del Bocco	Santa Margherita Staffora (PV)	*Quercus cerris*	supratemperate - humid mesotemperate
G.adsp.8	*Ganoderma adspersum*	(Schulzer) Donk	Morina	Rovescala	*Quercus petraea*	humid supratemperate - subhumid
G.adsp.9	*Ganoderma adspersum*	(Schulzer) Donk	RNIS Bosco Siro Negri	Zerbolò (PV)	*Quercus robur*	humid supratemperate - subhumid
G.adsp.10	*Ganoderma adspersum*	(Schulzer) Donk	RNIS Bosco Siro Negri	Zerbolò (PV)	*Populus nigra*	humid supratemperate - subhumid
G.adsp.11	*Ganoderma adspersum*	(Schulzer) Donk	RNIS Bosco Siro Negri	Zerbolò (PV)	*Quercus robur*	humid supratemperate - subhumid
G.adsp.12	*Ganoderma adspersum*	(Schulzer) Donk	Ca' del Bosco	Nibbiano Val Tidone (PC)	*Quercus cerris*	humid supratemperate - hyperhumid
G.adsp.13	*Ganoderma adspersum*	(Schulzer) Donk	Montebolone	Pavia (PV)	*Cedrus atlantica*	humid supratemperate - subhumid
G.adsp.14	*Ganoderma adspersum*	(Schulzer) Donk	Cascina Scova	Pavia (PV)	*Unidentified broadleaf*	humid supratemperate - subhumid
G.adsp.15	*Ganoderma adspersum*	(Schulzer) Donk	Ticino	Torre d'Isola (PV)	*Quercus sp.*	humid supratemperate - subhumid
G.adsp.17	*Ganoderma adspersum*	(Schulzer) Donk	unknown	Bologna (BO)	*Unidentified broadleaf*	supratemperate/humid mesotemperate-subhumid
G.car.1	*Ganoderma carnosum*	Pat.	Foreste Casentinesi	Poppi (AR)	*Abies alba*	hyperhumid supratemperate/ultrahyperhumid

Table 2. *Cont.*

Mic UNIPV ID	Species	Authors	Locality	Municipality	Host	Phytoclimate Class
G.pf.1	*Ganoderma pfeifferi*	Bres.	Prati di Tivo	Pietracamela (TE)	*Fagus sylvatica*	supratemperate/hyperhumid mesotemperate/humid
H.e.1	*Hericium erinaceus*	(Bull.) Pers.	Colle Ciupi	Siena (SI)	*Quercus ilex*	subhumid mesotemperate/humid
H.e.2	*Hericium erinaceus*	(Bull.) Pers.	Castello di Belcaro	Siena (SI)	*Quercus ilex*	subhumid mesotemperate/humid
H.e.3	*Hericium erinaceus*	(Bull.) Pers.	Strada per Castello di Belcaro	Siena (SI)	*Quercus ilex*	subhumid mesotemperate/humid
H.e.4	*Hericium erinaceus*	(Bull.) Pers.	Strada per Castello di Belcaro	Siena (SI)	*Quercus ilex*	subhumid mesotemperate/humid
H.e.5	*Hericium erinaceus*	(Bull.) Pers.	Colle Val d'Elsa	Colle Val d'Elsa (SI)	*Quercus ilex*	subhumid mesotemperate/humid
I.t.1	*Inocutis tamaricis*	(Pat.) Fiasson and Niemelä	Apani	Brindisi (BR)	*Tamarix gallica*	thermomediterranean/mesomediterranean/dry inframediterranean/subhumid
I.t.2	*Inocutis tamaricis*	(Pat.) Fiasson and Niemelä	Ostia Lido	Roma (RM)	*Tamarix gallica*	subhumid mesomediterranean
L.s.1	*Laetiporus sulphureus*	(Bull.) Murrill	RNIS Bosco Siro Negri	Zerbolò (PV)	*Quercus robur*	humid supratemperate - subhumid
L.s.2	*Laetiporus sulphureus*	(Bull.) Murrill	RNIS Bosco Siro Negri	Zerbolò (PV)	*Quercus robur*	humid supratemperate - subhumid
L.s.3	*Laetiporus sulphureus*	(Bull.) Murrill	RNIS Bosco Siro Negri	Zerbolò (PV)	*Quercus robur*	humid supratemperate - subhumid
L.s.4	*Laetiporus sulphureus*	(Bull.) Murrill	RNIS Bosco Siro Negri	Zerbolò (PV)	*Quercus robur*	humid supratemperate - subhumid
L.s.5	*Laetiporus sulphureus*	(Bull.) Murrill	Pietragavina	Varzi (PV)	*Castanea sativa*	supratemperate - hyperhumid mesotemperate - humid
L.s.6	*Laetiporus sulphureus*	(Bull.) Murrill	Cono di Volo Malpensa	Gallarate (VA)	*Quercus sp.*	mesotemperate-humid supratemperate
L.w.1	*Cellulariella warnieri*	(Durieu and Mont.) Zmitr. and V. Malysheva	R.N. Bosco Giuseppe Negri	Pavia (PV)	*Quercus robur*	humid supratemperate - subhumid
L.w.2	*Cellulariella warnieri*	(Durieu and Mont.) Zmitr. and V. Malysheva	Bosco del Cecco	Santa Maria della Versa (PV)	*Ulmus minor*	humid supratemperate - subhumid
L.w.3	*Cellulariella warnieri*	(Durieu and Mont.) Zmitr. and V. Malysheva	Rio Marsinola-Fracion	Rovescala (PV)	*Populus nigra*	humid supratemperate - subhumid
L.w.4	*Cellulariella warnieri*	(Durieu and Mont.) Zmitr. and V. Malysheva	RNIS Bosco Siro Negri	Zerbolò (PV)	*Quercus robur*	humid supratemperate - subhumid
L.w.5	*Cellulariella warnieri*	(Durieu and Mont.) Zmitr. and V. Malysheva	RNIS Bosco Siro Negri	Zerbolò (PV)	*Robinia pseudoacacia*	humid supratemperate - subhumid
L.w.6	*Cellulariella warnieri*	(Durieu and Mont.) Zmitr. and V. Malysheva	Bosco Sforza nord-Rio Bardonezza	Ziano Piacentino (PC)	*Ulmus minor*	humid supratemperate - subhumid
P.f.1	*Perenniporia fraxinea*	(Bull.) Ryv.	R.N. Bosco Giuseppe Negri	Pavia (PV)	*Populus nigra*	humid supratemperate - subhumid
P.f.2	*Perenniporia fraxinea*	(Bull.) Ryv.	Via Scala	Pavia (PV)	*Celtis australis*	humid supratemperate - subhumid
P.f.3	*Perenniporia fraxinea*	(Bull.) Ryv.	Rio Bardonezza	Rovescala (PV)	*Salix alba*	humid supratemperate - subhumid

Table 2. *Cont.*

Mic UNIPV ID	Species	Authors	Locality	Municipality	Host	Phytoclimate Class
P.f.4	*Perenniporia fraxinea*	(Bull.) Ryv.	Via Ubaldo degli Ubaldi	Pavia (PV)	*Unidentified broadleaf*	humid supratemperate - subhumid
P.f.5	*Perenniporia fraxinea*	(Bull.) Ryv.	Via Borgo Calvenzano	Pavia (PV)	*Platanus x hispanica*	humid supratemperate - subhumid
P.f.6	*Perenniporia fraxinea*	(Bull.) Ryv.	Via Borgo Calvenzano	Pavia (PV)	*Platanus x hispanica*	humid supratemperate - subhumid
P.f.7	*Perenniporia fraxinea*	(Bull.) Ryv.	Bosco Giuseppe Negri	Pavia (PV)	*Populus nigra*	humid supratemperate - subhumid
P.f.8	*Perenniporia fraxinea*	(Bull.) Ryv.	Bosco Giuseppe Negri	Pavia (PV)	*Populus nigra*	humid supratemperate - subhumid
P.f.9	*Perenniporia fraxinea*	(Bull.) Ryv.	Cascina Venara	Zerbolò (PV)	*Populus alba*	humid supratemperate - subhumid
P.f.10	*Perenniporia fraxinea*	(Bull.) Ryv.	Pizzofreddo	Santa Maria della Versa (PV)	*Unidentified broadleaf*	mesomediterranean - humid thermotemperate - subhumid
P.f.11	*Perenniporia fraxinea*	(Bull.) Ryv.	Parco della Vernavola	Pavia (PV)	*Robinia pseudoacacia*	humid supratemperate - subhumid
P.f.12	*Perenniporia fraxinea*	(Bull.) Ryv.	Parco della Vernavola	Pavia (PV)	*Robinia pseudoacacia*	humid supratemperate - subhumid
P.f.13	*Perenniporia fraxinea*	(Bull.) Ryv.	Parco della Vernavola	Pavia (PV)	*Robinia pseudoacacia*	humid supratemperate - subhumid
P.f.14	*Perenniporia fraxinea*	(Bull.) Ryv.	Parco della Vernavola	Pavia (PV)	*Robinia pseudoacacia*	humid supratemperate - subhumid
P.f.15	*Perenniporia fraxinea*	(Bull.) Ryv.	Parco della Vernavola	Pavia (PV)	*Robinia pseudoacacia*	humid supratemperate - subhumid
P.f.16	*Perenniporia fraxinea*	(Bull.) Ryv.	Parco della Vernavola	Pavia (PV)	*Robinia pseudoacacia*	humid supratemperate - subhumid
P.f.17	*Perenniporia fraxinea*	(Bull.) Ryv.	Parco della Vernavola	Pavia (PV)	*Robinia pseudoacacia*	humid supratemperate - subhumid
P.f.18	*Perenniporia fraxinea*	(Bull.) Ryv.	Parco della Vernavola	Pavia (PV)	*Robinia pseudoacacia*	humid supratemperate - subhumid
P.f.19	*Perenniporia fraxinea*	(Bull.) Ryv.	Via D. Alighieri 25	Illasi (VR)	*Olea europaea*	humid supratemperate - subhumid
P.f.20	*Perenniporia fraxinea*	(Bull.) Ryv.	via Mascherpa	Castelvetro Piacentino (PC)	*Populus alba*	humid supratemperate - subhumid
P.f.21	*Perenniporia fraxinea*	(Bull.) Ryv.	Lungolago	Mergozzo (VCO)	*Robinia pseudoacacia*	humid supratemperate/hyperhumid
P.f.22	*Perenniporia fraxinea*	(Bull.) Ryv.	Cascina Scova	Pavia (PV)	*Robinia pseudoacacia*	humid supratemperate - subhumid
P.f.23	*Perenniporia fraxinea*	(Bull.) Ryv.	RNIS Bosco Siro Negri	Zerbolò (PV)	*Quercus robur*	humid supratemperate - subhumid
P.f.24	*Perenniporia fraxinea*	(Bull.) Ryv.	RNIS Bosco Siro Negri	Zerbolò (PV)	*Populus nigra*	humid supratemperate - subhumid
P.f.25	*Perenniporia fraxinea*	(Bull.) Ryv.	Santa Sofia	Torre d'Isola (PV)	*Quercus robur*	humid supratemperate - subhumid
P.f.26	*Perenniporia fraxinea*	(Bull.) Ryv.	Cono di Volo Malpensa	Gallarate (VA)	*Unidentified broadleaf*	mesotemperate - humid supratemperate
P.f.27	*Perenniporia fraxinea*	(Bull.) Ryv.	Viale Gorizia - Mura Spagnole	Pavia (PV)	*Celtis australis*	humid supratemperate - subhumid
P.f.28	*Perenniporia fraxinea*	(Bull.) Ryv.	Fossone - Bosco della Fame	Rovescala (PV)	*Populus nigra*	humid supratemperate - subhumid
P.m.1	*Perenniporia meridionalis*	Decock and Stalpers	R.N. Regionale Piramidi di Zone	Zone (BS)	*Quercus robur*	hyperhumid supratemperate/humid
P.m.2	*Perenniporia meridionalis*	Decock and Stalpers	Sormano	Castellina in Chianti (SI)	*Olea europaea*	supratemperate/humid mesotemperate/hyperhumid

Table 2. *Cont.*

Mic UNIPV ID	Species	Authors	Locality	Municipality	Host	Phytoclimate Class
P.och.1	*Perenniporia ochroleuca*	(Berk.) Ryvarden	Belcaro	Siena (SI)	*Quercus ilex*	subhumid mesotemperate/humid
P.och.2	*Perenniporia ochroleuca*	(Berk.) Ryvarden	Le Manie	Savona (SV)	*Quercus ilex*	mesomediterranean/dry-subhumid thermomediterranean
Ph.c.1	*Phellinus contiguus*	(Pers.) Pat.	Rio Bardonezza	Ziano Piacentino (PC)	*Robinia pseudoacacia*	humid supratemperate - subhumid
Ph.tor.1	*Phellinus torulosus*	(Pers.) Bourdot and Galzin	Ticino	Torre d'Isola (PV)	*Prunus avium*	humid supratemperate - subhumid
Pl.e.1	*Pleurotus eryngii*	(DC.) Quél.	Aidomaggiore	Aidomaggiore (OR)	*unidentified*	subhumid mesomediterranean
Pl.e.2	*Pleurotus eryngii*	(DC.) Quél.	Spadafora	Spadafora (ME)	*unidentified*	thermomediterranean/subhumid mesomediterranean
Punc.s.1	*Punctularia strigosozonata*	(Schwein.) P.H.B. Talbot	Bosco di Bauli'	Palazzolo Acreide (SR)	*Quercus sp.*	mesomediterranean/dry thermomediterranean-subhumid
Sp.p.1	*Spongipellis pachyodon*	(Pers.) Kotl. and Pouzar	Rio Marsinola-Fracion	Rovescala (PV)	*Prunus avium*	humid supratemperate - subhumid
Sp.p.2	*Spongipellis pachyodon*	(Pers.) Kotl. and Pouzar	Civezza	Imperia (IM)	*Quercus pubescens*	mesomediterranean/dry thermomediterranean-subhumid

The species reported in Table 2 represent about one in five of the comprehensive collection of wood decay species in MicUNIPV. The temperate region and Mediterranean region in Italy are reciprocally intersected and several phytoclimates are represented based on both thermal–pluviometrical parameters and floristic–vegetational ones. According to our field observations, this has the consequence that several species can be found in different phytoclimates and on different hosts, whereas a minor fraction is strictly related to one or few hosts.

Daedaleopsis confragosa and *D. tricolor* are easily distinguished by morphology; nevertheless, ITS sequences are important to discriminate the species strain. *D. tricolor* seems more common in Central and Southern Europe; in Italy, it has been reported in seven out of 20 regions [45]. Our strain (MicUNIPV D.tric.1) comes from the lower Apennines in Pavia Province; other field observations suggest that *Prunus avium* is the favourite host of *D. tricolor* in North and Central Italy. The strain has not been characterized yet, although pharmacological effects have been reported [46].

Daedalea quercina has been reported in 11/20 Italian regions [45]; as expected, all MicUNIPV strains were isolated from *Quercus* spp. Nevertheless, strain MicUNIPV D.q.1 efficiently colonized poplar wood chips and confirmed that this species is a typical brown rot agent [47].

Despite being apparently cosmopolitan, *Desarmillaria tabescens* is strictly related to *Quercus* in warm climates, where it behaves as a secondary pathogen [48]. *D. tabescens* has been reported in 16/20 Italian Regions [45]. Accordingly, our strain was isolated from roots of *Q. robur* in RNIS Bosco Siro Negri (Pavia, Italy), which is a significant, unmanaged residue of typical forest of the western Po Plain.

Fomitopsis iberica is a rare species, reported in three Italian regions [45]. All the strains in MicUNIPV were isolated close to Varese lakes, either on broadleaves or conifers.

Ganoderma is represented in MicUNIPV by seven species: *G. adspersum*, *G. applanatum*, *G. carnosum*, *G. pfeifferi*, *G. lucidum*, *G. resinaceum* and *G. valesiacum*. This genus has been intensely studied due to its wide range of secondary metabolites, including several bioactive compounds [49]. According to Ryvarden and Melo [6], *Ganoderma* is one of the most difficult genera to identify at species level. As reported in Table 2, we obtained strains of *G. adspersum*, confirming that is a southern species in Europe [50]. *G. carnosum* is usually located in the *G. lucidum* complex due to its morphological similarity, despite it showing clear differences in host relationship. Our strain was isolated from its type-locality in Italy, i.e., a forest of *Abies alba*, that is likely to be its preferred host in South Europe [33]. Molecular identification by ITS region met difficulties in discriminating *G. carnosum* from *G. tsugae* and *G. oregonensis*; this topic presents questions about the real interspecific diversity within this conifer-related species in *Ganoderma*. Strains in pure culture will thus help us to investigate both the molecular and morphological nested diversity in this complex. An analogous problem concerns *G. pfeifferi*, as it partially shares its trophic niche with *G. lucidum* and *G. resinaceum*.

L. sulphureus is particularly related to *Quercus* according to our observations in North Italy, although *Castanea sativa* is also represented. Interestingly, *C. sativa* is also one of the favourite hosts of *Phellinus torulosus*, according to field observations, despite strain MicUNIPV Ph.tor.1 being isolated from *P. avium*.

Genus *Perenniporia* is represented in MicUNIPV by *P. fraxinea*, *P. meridionalis* and *P. ochroleuca*. According to our field observation, *P. fraxinea* is more common and widespread than expected, particularly in urban areas. We have focused our attention in indentifying strains, which, to date, number 27. Some of the isolated strains have been used for population studies and tests on heavy metal bioaccumulation [16,51].

Pleurotus eryngii is a typically Mediterranean species as well as its herbaceous hosts in *Apiaceae*. Consistently, the MicUNIPV Pl.e.1 and Pl.e.2 strains were isolated in properly Mediterranean areas (Sardinia and Sicily).

Punctularia strigosozonata is a rare, poorly studied species, typically related to the Mediterranean area; its resupinate morphology increases the difficulty in achieving pure isolates.

Spongipellis pachyodon has a mainly central–southern distribution in Europe; according to Onofri et al. [45], in Italy it is known in five out of 20 Regions, not including either Lombardy or Liguria.

This species is reported as uncommon but locally abundant; regarding this, our field observations suggest that the population in the Pavia-Piacenza Apennines is particularly related to *P. avium*.

Further species listed in Table 2 are reported below in more detail owing to their taxonomic controversy or potential applications.

3.1. *Cellulariella Warnieri (Durieu and Mont.) Zmitr. and V. Malysheva*

3.1.1. Background

As detailed in Table 2, the basidiomata of some species were collected in the Mediterranean area and others that are known to prefer warm environment, even if they were collected in continental or temperate zones. An example is *Cellulariella warnieri*, a poorly investigated species related to warm climates, according to Bernicchia and Gorjón [44]; despite not strictly being related to the Mediterranean region, Ryvarden and Melo [6] reported it as a southern and rare species.

The notable scarcity of data about this species has probably contributed to its uncertain systematic and taxonomic status. Currently it is reported as: *Lenzites warnieri* Durieu and Mont. by Mycobank [40], *C. warnieri* by Index Fungorum [52] and *Trametes warnieri* (Durieu and Mont.) Zmitr., Wasser and Ezhov by Ryvarden and Melo [6]. The latter indication is suggested also by Justo and Hibbett [53] based on a five marker-based phylogenetic classification of *Trametes*. Significantly, only 108 records for this species have been reported by the GBIF (Global Biodiversity Information Facility) [54] and only 10 sequences are available in GenBank, almost half of them being critical as they are reported from South East Asia [55]. Further analyses on a more representative number of strains are thus needed to clarify the position of *C. warnieri*.

Strain MicUNIPV L.w.1 was tested for the evaluation of lignocellulolytic activity and resulted in a very low production of Mn peroxydase and lack of lignin peroxydase, whereas cellulase and hemicellulase had the highest presentation among the species under examination [47]. This was also confirmed when testing the effect of its colonization on *Medicago sativa* for pre-treatment, as cellulose and hemicellulose were preferentially removed [56].

3.1.2. MicUNIPV WDF Strains Results

According to our field observations, localities are distributed as small local clusters which are very scattered in turn. Thus, the strains MicUNIPV L.w.1, L.w.4 and L.w.5 were collected from Po plain areas (Pavia and RNIS Bosco Siro Negri), whereas strains the MicUNIPV L.w.2, L.w.3 and L.w.6 were collected from hill area (Oltrepo Pavese).

The six strains of MicUNIPV were collected from *Quercus*, *Ulmus*, *Populus* and *Robinia*; interestingly, our field observations pointed out some preference for *Ulmus*, which was not previously reported as a host in Italy. As expected, all the strains were isolated from individuals behaving as saprotrophs. It should be noted that, since *C. warnieri* develops basidiomata in late autumn but it releases spores in spring, the basidioma itself remains vital even at low temperatures and under the snow.

3.2. *Dichomitus squalens (P. Karst.) D.A. Reid*

3.2.1. Background

Dichomitus squalens is a model species for studies about the selectivity of white rot and its enzymatic basis [57–60]. Despite being reported all throughout the boreal emisphere, it appears scattered and is commonly found in the northern parts of Europe, North America and Asia [61]; the GBIF [54] places the wide majority of records in the Fennoscandian region. The host relationship is apparently controversial and surprising: Ryvarden and Melo [6] assumed *Pinus* as the only European host species, whereas Bernicchia and Gorjón [44] recorded *Picea abies* for the Italian sample and Niemelä [62] assigned most samples to *Pinus* and a smaller fraction to *Picea abies* in Białowieża Forest (Poland / Belarus). Nevertheless, it should be noted that American samples have been reported on six different genera in *Pinaceae*. Furthermore, young basidiomata of *D. squalens* are easily misidentified due to the close morphological resemblance with *Neoantrodia serialis* (Fr.) Audet and related species. Consistently, a remarkable intraspecific variability in growth and enzyme profiles was revealed by testing different monokarya strains [61]. This is also consistent with the numerous mating types deriving from tetrapolarity [61].

3.2.2. MicUNIPV WDF Strains Results

Strains MicUNIPV D.sq.1 and MicUNIPV D.sq.2 were recovered from *Pinus pinea* and *Cedrus* sp. respectively near the Adriatic Sea and Varese Lake. The Italian strains have not yet been investigated for their enzymatic properties, so they may provide an additional tool to explore the diversity in degradation potential of this selective decayer.

3.3. *Hericium Erinaceus (Bull.) Pers.*

3.3.1. Background

Hericium erinaceus (Bull.) Pers. is one of the most famous cultivated medicinal species in the world; a wide range of peculiar compounds, both related to primary (e.g., β-glucans) and secondary metabolism (e.g., erinacines and hericenones) have been up to now characterized and screened for bioactivity [7,63–66]. GBIF [54] places the wide majority of *H. erinaceus* sites in Europe, North America and North Eastern Asia. According to the phylogenetic study by Cesaroni et al. [67], a subclade containing European and American ITS sequences is well distinguished from the Asian clade. Despite relying on ITS region only, these data suggest the possibility to differentiate *H. erinaceus* strains also by the phylogeographic structure. Notwithstanding the scarcity of available data for Asian samples, *H. erinaceus* apparently has a quite broad trophic niche including several host species in *Fagaceae,* and *Aceraceae* to lesser extent, and particularly showing a preference for *Quercus* all throughout its distribution area [37]. Consistently, European samples have mainly been recovered from *Quercus* and *Fagus*, the former likely being the exclusive host in Italy and the only known host in North Africa [34,68]. Strain MicUNIPV H.e.2 was analyzed for the production of erinacine A and hericenes (presumably A, B, C, D). Thus, a complete quali-quantitative comparison of these selected metabolites was provided throughout different growth stages but within the same strain, which is a powerful tool for the standardization of bioactive products [69].

Strain MicUNIPV H.e.1 was selected to test the effect of oral supplementation on mice memory. The results indicate an improvement in recognition memory and induction of hyppocampal and cerebellar neurogenesis during aging. This strain has therefore contributed to pointing out which areas are directly involved in the neuroactivity of *H. erinaceus* compounds, highlighting which type of memory is increased [70].

3.3.2. MicUNIPV WDF Strains Results

Accordingly, all four strains in the MicUNIPV collection were recovered in the municipality of Siena from *Q. ilex*, that is, a featuring species in the flora of Mediterranean area often forming homogeneous woodlands. It is noteworthy that the Mediterranean basin hosts a great variety of *Quercus* species, whose phylogenetic and systematic relationships are still controversial, with particular concern to the *Q. ilex* group [71–73].

3.4. *Inocutis Tamaricis (Pat.) Fiasson and Niemelä*

3.4.1. Background

The relationship of *Inocutis tamaricis* with *Tamarix* is apparently so strict to be regarded as a discriminant character in identification [6,33,44]. Although the genus *Tamarix* consists of 72 accepted species in Europe, Asia and Africa [74], *I. tamaricis* is restricted to the Mediterranean basin and

Macaronesia. Here, it grows on different *Tamarix* species according to their availability but shows a preference for *T. gallica* [75,76]. Consistently, strains MicUNIPV I.t.1 and I.t.2 were both isolated from *T. gallica*. As a whole, the intra-familiar phylogeny of *Hymenochaetaceae* is still to be clarified; multiple revisions have tried to point out nested diversity within polyphyletic taxa, such as *Inonotus* [77,78]. Thus, the genus *Inocutis* is nowadays accepted to be distinct from *Inonotus* itself, as formerly suggested by Fiasson and Niemelä [79]. Interestingly, the type-species for *Inocutis* is *I. rheades* (Pers.) Fiasson and Niemelä, which is morphologically very similar to *I. tamaricis* and is mostly distinguished by host and distribution [6]. Thus, *I. tamaricis* may be regarded as the Mediterranean counterpart of *I. rheades*. As a whole, only 12 sequences have been up to now deposited in GenBank [55] as belonging to this species, some of which are lacking data to assess their effective reliability. Further sequences from the Mediterranean area, equipped with information about host and geographic origin, are needed to support studies about intrageneric diversity in *Inocutis* in the light of biogeographic patterns.

3.4.2. MicUNIPV WDF Strains Results

Strains MicUNIPV I.t.1 and I.t.2 were both isolated from *T. gallica*, forming in both cases ornamental rows along the sea coast.

3.5. *Fomitiporia Mediterranea M. Fisch.*

3.5.1. Background

As mentioned for *Inonotus*, the genus *Phellinus* is increasingly revealing its hidden diversity; recognized as being polyphyletic, several species have been distributed into other genera, such as *Fomitiporia* Murrill. *Fomitiporia mediterranea* is a peculiar example due to its morphology, being actually indistinguishable from *P. punctatus*. According to Fischer [80], these two species also show differences in growth rate at selected temperatures and mating behaviour. The same study provides strong evidence for dichotomy in host selection by *F. mediterranea* depending on biogeography, i.e., this species grows on several tree species in Italy [81,82], whereas north of the Alps it apparently grows on *Vitis vinifera* exclusively [83,84].

It should be considered that misidentification with *P. punctatus* has probably led to the underestimation of *F. mediterranea* in the Mediterranean area [85]. Analogously, Polemis et al. [86] suggested that the relationship with *P. pseudopunctatus* A. David, Dequatre and Fiasson should be reconsidered as well, enclosing the latter in *F. mediterranea* clade. It may be observed that the *P. pseudopunctatus* is apparently more related to the South Mediterranean region and climates [87,88]. Further analyses on strains from different geographic origins and hosts are thus needed to clarify both phylogenetic relationships and biogeographic patterns. As a whole, this species complex is characterized by intense necrotrophic white rot; *F. mediterranea* in particular is regarded as one of the main agents responsible for wood rot in *V. vinifera*, *Corylus avellana* and *Olea europaea* [82,83,85].

3.5.2. MicUNIPV WDF Strains Results

All of the 17 Italian strains up to now attained by the Laboratory of Mycology DSTA–University of Pavia were assigned to *F. mediterranea* instead of *P. punctatus* and recovered from different substrates in North Italy. Even within one province (Pavia), *F. mediterranea* was located on five hosts, namely *Q. robur, Hedera helix* (State Natural Strict Reserve Bosco Siro Negri), *Salix alba* (University of Pavia courtyard), *R. pseudocacacia* and *V. vinifera* (Oltrepo Pavese hills).

The identified strains thus provide a tool to deepen pathology dynamics and different susceptibility depending on host species and cultivar.

3.6. *Perenniporia meridionalis Decock and Stalpers*

3.6.1. Background

Genus *Perenniporia* Murrill sensu lato is large, cosmopolitan and supposed to be polyphyletic, and thus is in need of further phylogenetic analysis based on sequences from different species [6]. In turn, an example of intrageneric complexity is provided by *P. meridionalis*, within *P. medulla-panis* (Jacq.) Donk group. Actually, the complete revision by Decock and Stalpers [89] arose the doubt that several records, as well as specimens in herbaria, are to be referred to *P. meridionalis* instead of *P. medulla-panis* (or closely related species), particularly when coming from the Mediterranean area. Up to now, only a partial and fragmentary investigation into the intra-generic diversity in *Perenniporia* has been carried out from a molecular and phylogenetic point of view [90]. It is noteworthy that no sequences at all are available in GenBank by the name of *P. meridionalis*, whereas 40 sequences are referred to *P. medulla-panis* [55]. Strains from culture collections are thus needed as basic material for this purpose. According to the indications by both Bernicchia and Gorjón [44], as well as Ryvarden and Melo [6], *P. meridionalis* is particularly related to *Quercus* (more than *P. medulla-panis*), the holotype having been isolated from *Q. ilex* in Sardinia [62,89].

3.6.2. MicUNIPV WDF Strains Results

Strains MicUNIPV P.m.1 and P.m.2 strains were respectively isolated from *Q. robur* (North Italy, near a lake) and *Q. ilex* (Central Italy). Besides considerations on biodiversity, *P. meridionalis* has a great applicative potential. MicUNIPV P.m.1 showed remarkable selectivity as a white rot agent and versatility when inoculated onto unusual substrates such as grass. The selective removal of lignin by this species contemporarily relies on high activity for Mn peroxidases and very low for one cellulase; the final delignification in the substrate is clear both in thermogravimetric analysis and FTIR spectroscopy [12,47].

3.7. *Perenniporia ochroleuca (Berk.) Ryvarden*

3.7.1. Background

Perenniporia ochroleuca is another example of the unsolved intra-generic diversity within *Perenniporia*. This species is suspected to hide a complex, and transfer to *Truncospora* Pilát ex Pilát has thus been suggested [91]. According to the same authors, the Iberian/Macaronesian clade gives *T. atlantica* Spirin and Vlasák, whereas the status of Australian samples is more uncertain, which would mainly belong to *T. ochroleuca*. Nevertheless, the new taxonomy has not yet been fully accepted, neither by Mycobank [40] nor by Index Fungorum [52]. These hypotheses therefore need to be supported by entering further sequences into the phylogenetic analyses from an exhaustive geographic range. *P. ochroleuca* was reported by Bernicchia and Gorjón [44] and Ryvarden and Melo [6] as tropical and growing on several hosts, whereas in Europe it is particularly related to the Mediterranean area. Nevertheless, Bernicchia and Gorjón [44] report a range of typically Mediterranean hosts, whereas Ryvarden and Melo [6] also include host plants whose distribution exceeds the Mediterranean area to include samples from the coasts of South England and Wales. Further phylogenetic analyses focused on the Mediterranean region versus the adjacent Atlantic ones are needed to test the monophyly of the proposed *T. atlantica*.

3.7.2. MicUNIPV WDF Strains Results

Both strains MicUNIPV P.och.1 and P.och.2 were isolated from *Q. ilex* in Central Italy and the Ligurian west coast, respectively.

4. Conclusions

At the moment, MicUNIPV, the fungal research culture collection of University of Pavia (Italy), maintains 500 strains from wood decay species. Examples particularly correlated to the Mediterranean area were discussed and their roles in accomplished research were mentioned in this study.

Culture collections of wood decay fungi are an important tool both for systematic and applied studies. Strains in pure culture are more easily and reliably identified and analyzed for metabolic activities and competitivity. The environmental features of the strain origin place have often been underestimated; nevertheless, the diversity of wood decay fungi strongly depends on biogeography and is related to host distribution. This also highlights the need for an investigation including a wider concept of the Mediterranean region than one strictly limited by climate or phytoclimate classification, i.e., even continental regions surrounding the Mediterranean area contribute to the explanation of Mediterranean diversity.

The Laboratory of Mycology DSTA–University of Pavia (Italy) has up to now successfully collaborated with both researchers from other universities and amateurs in order to increase the diversity richness and geographic origin range of strains, as well as to enter these strains in original pure and applied research such as MATER and CE4WE (grants from Cariplo Foundation and Regione Lombardia).

Author Contributions: Conceptualization, E.S. and C.E.G.; Methodology, E.S., A.B., A.M.P., C.E.G., R.M.B. and M.C.; Investigation, E.S., A.B., C.E.G., R.M.B., S.B. and M.C. Resources, E.S., A.M.P. and F.B.; Data Curation, E.S., A.B. and C.E.G.; Writing – Original Draft Preparation, C.E.G., S.B.; Writing – Review and Editing, C.E.G., E.S., A.B., A.M.P.; Supervision, E.S., A.M.P.; Project Administration, E.S.; Funding Acquisition, E.S., F.B. All authors have read and agreed to the published version of the manuscript.

Funding: This project has been funded by: Fondo Ricerca e Giovani dell'Università degli Studi di Pavia (Savino 2019); Fondazione Cariplo, grant n° 2018-1765 entitled "Myco-advanced leather materials (MATER)".

Acknowledgments: The State Natural Strict Reserve (RNIS) Bosco Siro Negri – Zerbolò for support and authorization for sampling; the Botanical Garden of Pavia (Orto Botanico di Pavia) for authorization for sampling. The authors are also grateful to C. Perini, University of Siena (Italy), Valentina Cesaroni and Fabio Savino for providing some original specimens.

Conflicts of Interest: The authors declare no conflict of interest.

References

1. Liers, C.; Arnstadt, T.; Ullrich, R.; Hofrichter, M. Patterns of lignin degradation and oxidative enzyme secretion by different wood-and litter-colonizing basidiomycetes and ascomycetes grown on beech-wood. *FEMS Microbiol. Ecol.* **2011**, *78*, 91–102. [CrossRef] [PubMed]
2. Van den Brink, J.; de Vries, R.P. Fungal enzyme sets for plant polysaccharide degradation. *Appl. Microbiol. Biotechnol.* **2011**, *91*, 1477. [CrossRef] [PubMed]
3. Riley, R.; Salamov A., A.; Brown, D.W.; Nagy, L.G.; Floudas, D.; Held, B.W.; Levasseur, A.; Lombard, V.; Morin, E.; Otillar, R.; et al. Extensive sampling of basidiomycete genomes demonstrates inadequacy of the white-rot/brown-rot paradigm for wood decay fungi. *Proc. Natl. Acad. Sci. USA* **2014**, *111*, 9923–9928. [CrossRef]
4. Hibbett, D.S.; Bauer, R.; Binder, M.; Giachini, A.J.; Hosaka, K.; Justo, A.; Larsson, E.; Larsson, K.H.; Lawrey, J.D.; Miettinen, O.; et al. Agaricomycetes. In *the Mycota*, 2nd ed.; McLaughlin, D.J., Spatafora, J.W., Eds.; Springer: Berlin, Germany, 2014; Volume 7, pp. 373–429. [CrossRef]
5. Nagy, L.G.; Riley, R.; Tritt, A.; Adam, C.; Daum, C.; Floudas, D.; Sun, H.; Yadav, J.S.; Pangilinan, J.; Larsson, K.H.; et al. Comparative genomics of early-diverging mushroom-forming fungi provides insights into the origins of lignocellulose decay capabilities. *Mol. Biol. Evol.* **2015**, *33*, 959–970. [CrossRef] [PubMed]
6. Ryvarden, L.; Melo, I. *Poroid Fungi of Europe*, 2nd ed.; Fungiflora: Oslo, Norway, 2017.
7. Ma, B.J.; Shen, J.W.; Yu, H.Y.; Ruan, Y.; Wu, T.T.; Zhao, X. Hericenones and erinacines: Stimulators of nerve growth factor (NGF) biosynthesis in Hericium erinaceus. *Mycology* **2010**, *1*, 92–98. [CrossRef]

8. Dresch, P.; Rosam, K.; Grienke, U.; Rollinger, J.M.; Peintner, U. Fungal strain matters: Colony growth and bioactivity of the European medicinal polypores Fomes fomentarius, Fomitopsis pinicola and Piptoporus betulinus. *AMB Express* **2015**, *5*, 4. [CrossRef]
9. Angelini, P.; Girometta, C.; Tirillini, B.; Moretti, S.; Covino, S.; Cipriani, M.; D'Ellena, E.; Angeles, G.; Federici, E.; Savino, E.; et al. A comparative study of the antimicrobial and antioxidant activities of Inonotus hispidus fruit and their mycelia extracts. *Int. J. Food Prop.* **2019**, *22*, 768–783. [CrossRef]
10. Pozdnyakova, N.N. Involvement of the ligninolytic system of white-rot and litter-decomposing fungi in the degradation of polycyclic aromatic hydrocarbons. *Biotechnol. Res. Int.* **2012**, *2012*. [CrossRef]
11. Giles, R.L.; Galloway, E.R.; Zackeru, J.C.; Naithani, V.; Parrow, M.W. Two stage fungal biopulping solubilizes lignocellulosic carbohydrates without supplemental enzymatic hydrolysis. *Int. Biodeter. Biodegradation* **2014**, *86*, 265–271. [CrossRef]
12. Girometta, C.; Zeffiro, A.; Malagodi, M.; Savino, E.; Doria, E.; Nielsen, E.; Buttafava, A.; Dondi, D. Pretreatment of alfalfa stems by wood decay fungus Perenniporia meridionalis improves cellulose degradation and minimizes the use of chemicals. *Cellulose* **2017**, *24*, 3803–3813. [CrossRef]
13. Mendonça Maciel, M.J.; Ribeiro, H.C.T. Industrial and biotechnological applications of ligninolytic enzymes of the basidiomycota: A review. *Electron. J. Biotechnol.* **2010**, *13*, 14–15. [CrossRef]
14. Gadd, G.M. Biosorption: Critical review of scientific rationale, environmental importance and significance for pollution treatment. *J. Chem. Technol. Biot.* **2009**, *84*, 13–28. [CrossRef]
15. Javaid, A.M.N.A.; Bajwa, R.; Manzoor, T. Biosorption of heavy metals by pretreated biomass of Aspergillus niger. *Pak. J. Bot.* **2011**, *43*, 419–425.
16. Sturini, M.; Girometta, C.; Maraschi, F.; Savino, E.; Profumo, A. A Preliminary Investigation on Metal Bioaccumulation by Perenniporia fraxinea. *B Environ. Contam. Tox* **2017**, *98*, 508–512. [CrossRef]
17. Arciniegas, A.; Prieto, F.; Brancheriau, L.; Lasaygues, P. Literature review of acoustic and ultrasonic tomography in standing trees. *Trees* **2014**, *28*, 1559–1567. [CrossRef]
18. CBS-KNAW Collections. Available online: www.cbs.knaw.nl (accessed on 29 December 2019).
19. All Russian Collection of Microorganisms—VKM. Available online: www.vkm.ru (accessed on 29 December 2019).
20. Baldrian, P.; Gabriel, J. Intraspecific variability in growth response to cadmium of the wood-rotting fungus Piptoporus betulinus. *Mycologia* **2002**, *94*, 428–436. [CrossRef]
21. Pawlik, A.; Janusz, G.; Dębska, I.; Siwulski, M.; Frąc, M.; Rogalski, J. Genetic and metabolic intraspecific biodiversity of Ganoderma lucidum. *BioMed Res. Int.* **2015**, *2015*. [CrossRef]
22. Zeng, Z.; Sun, H.; Vainio, E.J.; Raffaello, T.; Kovalchuk, A.; Morin, E.; Duplessis, S.; Asiegbu, F.O. Intraspecific comparative genomics of isolates of the Norway spruce pathogen (Heterobasidion parviporum) and identification of its potential virulence factors. *BMC Genom.* **2018**, *19*, 220. [CrossRef]
23. WFCC—World Federation for Culture Collections. Available online: www.wfcc.info (accessed on 29 December 2019).
24. MIRRI—Microbial Resorurce Research Infrastructure. Available online: www.mirri.org (accessed on 29 December 2019).
25. Gargano, M.L. Mycotheca of edible and medicinal mushrooms at herbarium SAF as a potential source of nutraceuticals and cultivated mushrooms. *Int. J. Med. Mushrooms* **2018**, *20*, 405–409. [CrossRef]
26. Caretta, G. Micologo a Pavia. In *Raffaele Ciferri Scienziato Versatile e Critico, Cisalpino*; Istituto Editoriale Universitario: Milano, Italy, 2000; pp. 51–54.
27. Lowry, D.B. Ecotypes and the controversy over stages in the formation of new species. *Biol. J. Linn. Soc.* **2012**, *106*, 241–257. [CrossRef]
28. Angelini, P.; Compagno, R.; Arcangeli, A.; Bistocchi, G.; Gargano, M.L.; Venanzoni, R.; Venturella, G. Macrofungal diversity and ecology in two Mediterranean forest ecosystems. *Plant Biosyst.* **2016**, *150*, 540–549. [CrossRef]
29. Ministero Dell'ambiente e Della Tutela del Territorio e del Mare. Geoportale Nazionale—Nuovo Visualizzatore. Available online: http://www.pcn.minambiente.it/viewer/ (accessed on 28 December 2019).
30. Rivas-Martínez, S.; Penas, Á.; Díaz, T.E. Bioclimatic and Biogeographic Maps of Europe. 1:16.000.000 University of León, E-24071, Spain. 2004. Available online: http://www.globalbioclimatics.org/form/bg_mEd.htm (accessed on 28 December 2019).

31. Blasi, C.; Capotorti, G.; Copiz, R.; Guida, D.; Mollo, B.; Smiraglia, D.; Zavattero, L. Classification and mapping of the ecoregions of Italy. *Plant Biosyst.* **2014**, *148*, 1255–1345. [CrossRef]
32. Sutherland, W.J. *Ecological Census Techniques: A Handbook*, 2nd ed.; Cambridge University Press: Cambridge, UK, 2006.
33. Bernicchia, A. *Polyporaceae sl.*; Candusso: Alassio (SV), Italy, 2005.
34. Bernicchia, A.; Gorjón, S.P. *Corticiaceae sl.*; Candusso: Alassio (SV), Italy, 2010.
35. Stalpers, J.A. Identification of wood-inhabiting fungi in pure culture. *Stud. Mycol.* **1978**, *16*, 1–248.
36. Gams, W.; Hoekstra, E.S.; Aptroot, A. *CBS Course of Mycology*; Centraalbureau voor Schimmelcultures Baarn: Delft, NL, USA, 1998.
37. Tamets, P. *Growing Gourmet and Medicinal Mushrooms*; Ten Speed Press: Berkeley, CA, USA, 2011.
38. Toju, H.; Tanabe, A.S.; Yamamoto, S.; Sato, H. High-coverage ITS primers for the DNA-based identification of ascomycetes and basidiomycetes in environmental samples. *PLoS ONE* **2012**, *7*, e40863. [CrossRef]
39. Nilsson, R.H.; Hyde, K.D.; Pawlowska, J.; Ryberg, M.; Tedersoo, L.; Aas, A.B.; Alias, S.A.; Alves, A.; Anderson, C.L.; Antonelli, A.; et al. Improving ITS sequence data for identification of plant pathogenic fungi. *Fungal Divers* **2014**, *67*, 11–19. [CrossRef]
40. Mycobank. Available online: www.mycobank.org (accessed on 19 December 2019).
41. Altobelli, E.; Bernicchia, A.; Pecoraro, L.; Savino, E. Raccolta, isolamento e coltivazione di funghi poliporoidi con proprietà medicinali. *Micol. Ital.* **2012**, *41*, 3–10.
42. Savino, E.; Girometta, C.; Chinaglia, S.; Guglielminetti, M.; Rodolfi, M.; Bernicchia, A.; Perini, C.; Salerni, E.; Picco, A.M. Medicinal mushrooms in Italy and their ex situ conservation through culture collection. In Proceedings of the 8th International Conference on Mushroom Biology and Mushroom Products, New Delhi, India, 19–22 Novemver 2014; pp. 50–54.
43. Savino, E.; Girometta, C.; Miteva-Staleva, J.; Kostadinova, A.; Krumova, E. Wood decay macrofungi: Strain collection and studies about antioxidant properties. *Comptes Rendus l'Académie Bulg. Sci.* **2016**, *69*, 747–755.
44. Bernicchia, A.; Gorjón, S.P. *Polypores of Mediterranean Region*; S.P. Publishers: Bologna, Italy, 2019; in press.
45. Onofri, S.; Bernicchia, A.; Filipello Marchisio, V.; Padovan, F.; Perini, C.; Ripa, C.; Salerni, E.; Savino, E.; Venturella, G.; Vizzini, A.; et al. *Checklist dei Funghi Italiani Basidiomycetes, Basidiomycota*, 1st ed.; Carlo Delfino Editore: Sassari, Italy, 2005; pp. 1–380.
46. Bernicchia, A.; Fugazzola, M.A.; Gemelli, V.; Mantovani, B.; Lucchetti, A.; Cesari, M.; Speroni, E. DNA recovered and sequenced from an almost 7000 y-old Neolithic polypore, Daedaleopsis tricolor. *Mycol. Res.* **2006**, *110*, 14–17. [CrossRef]
47. Doria, E.; Altobelli, E.; Girometta, C.; Nielsen, E.; Zhang, T.; Savino, E. Evaluation of lignocellulolytic activities of ten fungal species able to degrade poplar wood. *Int. Biodeter Biodegrad.* **2014**, *94*, 160–166. [CrossRef]
48. Koch, R.A.; Wilson, A.W.; Séné, O.; Henkel, T.W.; Aime, M.C. Resolved phylogeny and biogeography of the root pathogen Armillaria and its gasteroid relative, Guyanagaster. *BMC Evol. Biol.* **2017**, *17*, 33. [CrossRef]
49. Lindequist, U.; Jülich, W.D.; Witt, S. Ganoderma pfeifferi—A European relative of Ganoderma lucidum. *Phytochemistry* **2015**, *114*, 102–108. [CrossRef] [PubMed]
50. Karadelev, M.; Rusevska, K.; Kajevska, I. Distribution and ecology of Genus Ganoderma (Ganodermataceae) in the Republic of Macedonia. In Proceedings of the International Conference on Biological and Enveronmental Sciences, Tirana, Albania, 26–28 September 2008; pp. 320–326.
51. Sillo, F.; Savino, E.; Giordano, L.; Girometta, C.; Astegiano, D.; Picco, A.M.; Gonthier, P. Analysis of genotypic diversity provides a first glimpse on the patterns of spread of the wood decay fungus Perenniporla fraxinea in an urban park in northern Italy. *J. Plant Pathol.* **2016**, *98*, 617–624.
52. Index Fungorum. Available online: www.indexfungorum (accessed on 19 December 2019).
53. Justo, A.; Hibbett, D.S. Phylogenetic classification of Trametes (Basidiomycota, Polyporales) based on a five-marker dataset. *Taxon* **2011**, *60*, 1567–1583. [CrossRef]
54. GBIF—Global Biodiversity Information Facility. Available online: www.gbif.org (accessed on 28 December 2019).
55. GenBank- NCBI (National Center for Biotechnology Information). Available online: https://www.ncbi.nlm.nih.gov (accessed on 19 December 2019).

56. Zeffiro, A.; Dondi, D.; Marconi, R.P.; Malagodi, M.; Girometta, C.; Bentivoglio, A.; Lazzaroni, S.; Savino, E.; Nielsen, E.; Buttafava, A. Sugar Production for bioethanol from alfalfa stems. Results, and comparative study with application of lignocellulolytic activities of novel fungal species. In Atti del Convegno, Proceedings of the VI Workshop nazionale AICIng "Molecules and materials: Chemistry for engineering", Roma, Italy, 22–23 June 2015.
57. Floudas, D.; Binder, M.; Riley, R.; Barry, K.; Blanchette, R.A.; Henrissat, B.; Martínez, A.T.; Otillar, R.; Spatafora, J.W. The Paleozoic origin of enzymatic lignin decomposition reconstructed from 31 fungal genomes. *Science* **2012**, *336*, 1715–1719. [CrossRef] [PubMed]
58. Rytioja, J.; Hildén, K.; Di Falco, M.; Zhou, M.; Aguilar-Pontes, M.V.; Sietiö, O.M.; Tsang, A.; de Vries, R.P.; Mäkelä, M.R. The molecular response of the white-rot fungus Dichomitus squalens to wood and non-woody biomass as examined by transcriptome and exoproteome analyses. *Environ. Microbiol.* **2017**, *19*, 1237–1250. [CrossRef] [PubMed]
59. Daly, P.; López, S.C.; Peng, M.; Lancefield, C.S.; Purvine, S.O.; Kim, Y.M.; Zink, E.M.; Dohnalkova, A.; Singan, V.R.; Lipzen, A.; et al. Dichomitus squalens partially tailors its molecular responses to the composition of solid wood. *Environ. Microbiol.* **2018**, *20*, 4141–4156. [CrossRef]
60. López, S.C.; Peng, M.; Issak, T.Y.; Daly, P.; de Vries, R.P.; Mäkelä, M.R. Induction of genes encoding plant cell wall-degrading carbohydrate-active enzymes by lignocellulose-derived monosaccharides and cellobiose in the white-rot fungus Dichomitus squalens. *Appl. Environ. Microbiol.* **2018**, *84*, e00403-18. [CrossRef]
61. L'opez, S.C.; Theelen, B.; Manserra, S.; Issak, T.Y.; Rytioja, J.; Mäkelä, M.R.; de Vries, R.P. Functional diversity in Dichomitus squalens monokaryons. *IMA Fungus* **2017**, *8*, 17. [CrossRef]
62. Niemelä, T. *Polypores of the Białowieża Forest*; Białowieski Park Narodowy: Białowieża, Poland, 2013.
63. Kawagishi, H.; Shimada, A.; Shirai, R.; Okamoto, K.; Ojima, F.; Sakamoto, H.; Ishiguro, Y.; Furukawa, S. Erinacines A, B and C, strong stimulators of nerve growth factor (NGF)-synthesis, from the mycelia of Hericium erinaceum. *Tetrahedron Lett.* **1994**, *35*, 1569–1572. [CrossRef]
64. Friedman, M. Chemistry, nutrition, and health-promoting properties of Hericium erinaceus (lion's mane) mushroom fruiting bodies and mycelia and their bioactive compounds. *J. Agric. Food Chem* **2015**, *63*, 7108–7123. [CrossRef]
65. Brandalise, F.; Cesaroni, V.; Gregori, A.; Repetti, M.; Romano, C.; Orrù, G.; Botta, L.; Girometta, C.; Guglielminetti, M.L.; Savino, E.; et al. Dietary supplementation of Hericium erinaceus increases mossy fiber-CA3 hippocampal neurotransmission and recognition memory in wild-type mice. *Evid-Based Compl. Alt.* **2017**. [CrossRef] [PubMed]
66. Rossi, P.; Cesaroni, V.; Brandalise, F.; Occhinegro, A.; Ratto, D.; Perrucci, F.; Lanaia, V.; Girometta, C.; Orrù, G.; Savino, E. Dietary supplementation of lion's mane medicinal mushroom, Hericium erinaceus (Agaricomycetes), and spatial memory in wild-type mice. *Int. J. Med. Mushrooms* **2018**, *20*, 485–494. [CrossRef] [PubMed]
67. Cesaroni, V.; Brusoni, M.; Cusaro, C.M.; Girometta, C.; Perini, C.; Picco, A.M.; Rossi, P.; Salerni, E.; Savino, E. Phylogenetic Comparison between Italian and Worldwide Hericium Species (Agaricomycetes). *Int. J. Med. Mushrooms* **2019**, *21*, 943–954. [CrossRef]
68. Ouali, Z.; Sbissi, I.; Boudagga, S.; Rhaiem, A.; Hamdi, C.; Venturella, G.; Saporita, P.; Jaouani, A.; Gargano, M.L. First report of the rare tooth fungus Hericium erinaceus in North African temperate forests. *Plant Biosyst.* **2018**, *154*, 24–28. [CrossRef]
69. Corana, F.; Cesaroni, V.; Mannucci, B.; Baiguera, R.M.; Picco, A.M.; Savino, E.; Ratto, D.; Perini, C.; Kawagishi, H.; Girometta, C.E.; et al. Array of Metabolites in Italian Hericium erinaceus Mycelium, Primordium, and Sporophore. *Molecules* **2019**, *24*, 3511. [CrossRef] [PubMed]
70. Ratto, D.; Corana, F.; Mannucci, B.; Priori, E.C.; Cobelli, F.; Roda, E.; Ferrari, B.; Occhinegro, A.; Di Iorio, C.; De Luca, F.; et al. Hericium erinaceus Improves recognition memory and induces hippocampal and cerebellar neurogenesis in frail mice during aging. *Nutrients* **2019**, *11*, 715. [CrossRef]
71. Manos, P.S.; Doyle, J.J.; Nixon, K.C. Phylogeny, biogeography, and processes of molecular differentiation in Quercus subgenus Quercus (Fagaceae). *Mol. Phylogenet. Evol.* **1999**, *12*, 333–349. [CrossRef]
72. Simeone, M.C.; Piredda, R.; Papini, A.; Vessella, F.; Schirone, B. Application of plastid and nuclear markers to DNA barcoding of Euro-Mediterranean oaks (Quercus, Fagaceae): Problems, prospects and phylogenetic implications. *Bot. J. Linn. Soc.* **2013**, *172*, 478–499. [CrossRef]

73. Hubert, F.; Grimm, G.W.; Jousselin, E.; Berry, V.; Franc, A.; Kremer, A. Multiple nuclear genes stabilize the phylogenetic backbone of the genus Quercus. *Syst. Biodivers.* **2014**, *12*, 405–423. [CrossRef]
74. Kew Science—Plants of the World Online. Available online: http://www.plantsoftheworldonline.org/ (accessed on 19 December 2019).
75. Loizides, M. Diversity of wood-inhabiting aphyllophoraceous basidiomycetes on the island of Cyprus. *Mycotaxon* **2017**, *132*, 985–986.
76. Gargano, M.L. Nuovi ospiti di Inonotus tamaricis (Hymenochaetaceae) in Sicilia. *Inf. Bot. Ital.* **2010**, *42*, 319–321.
77. Wagner, T.; Fischer, M. Natural groups and a revised system for the European poroid Hymenochaetales (Basidiomycota) supported by nLSU rDNA sequence data. *Mycol. Res.* **2001**, *105*, 773–782. [CrossRef]
78. Wagner, T.; Fischer, M. Proceedings towards a natural classification of the worldwide taxa Phellinus s.l. and Inonotus s.l., and phylogenetic relationships of allied genera. *Mycologia* **2002**, *94*, 998–1016. [CrossRef] [PubMed]
79. Fiasson, J.L.; Niemelä, T. The Hymenochaetales: A revision of the European poroid taxa. *Karstenia* **1984**, *24*, 14–28. [CrossRef]
80. Fischer, M. A new wood-decaying basidiomycete species associated with esca of grapevine: Fomitiporia mediterranea (Hymenochaetales). *Mycol. Prog.* **2002**, *1*, 315–324. [CrossRef]
81. Ciccarone, C.; Graniti, A.; Schiaffino, A.; Marras, F. Molecular analysis of Fomitiporia mediterranea isolates from esca-affected grapevines in southern Italy. *Phytopathol. Mediterr.* **2004**, *43*, 268–272.
82. Pilotti, M.; Tizzani, L.; Brunetti, A.; Gervasi, F.; Di Lernia, G.; Lumia, V. Molecular identification of Fomitiporia mediterranea on declining and decayed hazelnut. *JPP* **2010**, *92*, 115–129.
83. Fischer, M. Biodiversity and geographic distribution of basidiomycetes causing esca-associated white rot in grapevine: A worldwide perspective. *Phytopathol. Mediterr.* **2006**, *45*, 30–42.
84. Kovács, C.; Sándor, E. The increasing importance of grapevine trunk diseases. *Int. J. Hortic. Sci.* **2016**, *22*, 21–30. [CrossRef]
85. Markakis, E.A.; Ligoxigakis, E.K.; Roussos, P.A.; Sergentani, C.K.; Kavroulakis, N.; Roditakis, E.N.; Koubouris, G.C. Differential susceptibility responses of Greek olive cultivars to Fomitiporia mediterranea. *Eur. J. Plant Pathol.* **2019**, *153*, 1055–1066. [CrossRef]
86. Polemis, E.; Dimou, D.M.; Fryssouli, V.; Zervakis, G.I. Diversity of saproxylic basidiomycetes in Quercus ilex woodlands of central and insular Greece. *Plant Biosyst.* **2019**, *153*, 385–397. [CrossRef]
87. Saitta, A.; Venturella, G. On the presence of Diplomitoporus lindbladii and Phellinus pseudopunctatus in Sicily (Southern Italy). *Bocconea* **2009**, *23*, 273–276.
88. Karadelev, M.; Rusevska, K.; Kost, G.; Kopanja, D.M. Checklist of macrofungal species from the phylum Basidiomycota of the Republic of Macedonia. *Acta Musei Maced. Sci. Nat.* **2018**, *21*, 23–112.
89. Decock, C.; Stalpers, J.A. Studies in Perenniporia: Polyporus unitus, Boletus medulla-panis, the nomenclature of Perenniporia, Poria and Physisporus, and a note on European Perenniporia with a resupinate basidiome. *Taxon* **2006**, *55*, 759–778. [CrossRef]
90. Zhao, C.L.; Cui, B.K.; Dai, Y.C. New species and phylogeny of Perenniporia based on morphological and molecular characters. *Fungal Divers.* **2013**, *58*, 47–60. [CrossRef]
91. Spirin, V.; Kout, J.; Vlasák, J. Studies in the Truncospora ohiensis–T. ochroleuca group (Polyporales, Basidiomycota). *Nova Hedwig.* **2015**, *100*, 159–175. [CrossRef]

diversity

MDPI

Article

Does Arbuscular Mycorrhiza Determine Soil Microbial Functionality in Nutrient-Limited Mediterranean Arid Ecosystems?

Neji Mahmoudi [1], Teresa Dias [2], Mosbah Mahdhi [3], Cristina Cruz [2], Mohamed Mars [1] and Maria F. Caeiro [4,*]

1 Unité de Recherche Biodiversité et Valorisation des Bio-ressources en Zones Arides (BVBZA), Faculté des Sciences de Gabès, Cité Erriadh Zrig 6072, Tunisie; neji.2012@yahoo.com (N.M.); Mohamed.Mars@fsg.rnu.tn (M.M.)

2 Centre for Ecology, Evolution and Environmental Changes (cE3c), Faculdade de Ciências da Universidade de Lisboa, Edifício C2, Piso 5, Sala 2.5.03 Campo Grande, 1749-016 Lisboa, Portugal; mtdias@fc.ul.pt (T.D.); cmhoughton@fc.ul.pt (C.C.)

3 Center for Environmental Research and Studies, Jazan University, Jazan-P.O. Box 114, Jazan 82817, Saudi Arabia; mmahdhi@jazanu.edu.sa

4 Centro de Estudos do Ambiente e do Mar (CESAM), Faculdade de Ciências da Universidade de Lisboa, Edifício C2, Piso 2, Campo Grande, 1749-016 Lisboa, Portugal

* Correspondence: mfcaeiro@fc.ul.pt

Received: 5 May 2020; Accepted: 4 June 2020; Published: 10 June 2020

Abstract: Arbuscular mycorrhizal fungi (AMF) are determinant for the performance of plant communities and for the functionality of terrestrial ecosystems. In natural ecosystems, grazing can have a major impact on mycorrhizal fungi and consequently on plant growth. The objective of this study was to evaluate the statements referred above in Mediterranean arid areas in Tunisia. Root samples and rhizosphere soils of five dominant herbaceous plants were studied at six distinct arid sites differing on soil proprieties and grazing intensity. At each site, chemical and dynamic properties of the soil were characterized as well as the AMF colonization intensity and the soil functionality. Results showed that the mycorrhizal frequency and intensity and spore density, varied between plants in the same site and, for each plant, between sites and evidenced a positive effect of mycorrhized plants on soil microbial activity. Grazing and soil properties strongly affected AMF composition and the soil microbial and biochemical dynamics, which presented the lowest values at the sites with the highest grazing intensities. In conclusion, these results demonstrate that AMF improve soil biological properties, supporting the hypothesis that mycorrhiza and grazing compete for plant photosynthates, and highlight the importance of mycorrhizal symbiosis towards soil functionality under arid conditions.

Keywords: arbuscular mycorrhizal fungi; arid areas; biological properties; conserved areas; grazing; mycorrhiza

1. Introduction

Arid and semi-arid regions of the world are considered as being particularly vulnerable to climate change [1]. They are already climatically stressed with high temperatures, low rainfall and long dry seasons. These ecosystems are highly dynamic, with bursts of productivity in the wet season of some years, and very low productivity in dry years. The arid and semi-arid regions of the world have been subjected to accelerated desertification due to increasing grazing intensity, decreased rainfall, higher temperatures and prolonged periods of drought [2,3]. These pressures, associated with increased anthropogenic impacts and climate changes, caused the decline of forests, regression and extinction of many pastoral and forage species, and accelerated soil degradation and change of soil microbial communities [1].

In most cases, the degradation process starts with the disruption of functional networks that protect and alleviate stress, conferring to the organisms involved protection to global changes [4], including land use changes. Plants are well known for their symbioses with nitrogen fixing bacteria and arbuscular mycorrhiza fungi (AMF). These are just two examples of a vast range of symbiotic relationships that consist the plant microbiome and modulate plant phenotype and, thus, plant fitness. In this context, the microbial community of the rhizosphere [5] is of great importance to plant performance, playing a crucial role in ecosystem functioning. The microbial activities of the rhizosphere also determine the bioavailability of nutrients and, therefore, soil fertility. Plants interact with guilds of these beneficial microorganisms living in their rhizosphere, promoting their growth and development [6].

Arbuscular mycorrhiza involves reciprocal complementary resource exchanges between plant roots and soil fungi and is widespread in natural ecosystems [7]. When colonizing the roots, AMF develop intra- and extra-radical mycelium. The hyphae of the intra-radical mycelium colonize the cells of the root cortex and penetrate the periplasmic space where they develop vesicles and arbuscules that are the structures responsible for most of the exchanges between the fungi and the plant [8,9]. The extra-radical mycelium spreads its hyphae beyond the root surface and colonizes the surrounding substratum, increasing the volume of soil explored, and creating a privileged space for microbial development. As part of the interaction, the host plant provides the fungus with carbon in exchange for nutrients and water is taken up by the fungus [8,10]. Thus, mycorrhiza plays a crucial role in terrestrial ecosystem functioning, especially in arid or semi-arid areas, where root exudates are the major carbon source supporting soil microbial activities [11]. Apart from the nutritional benefits, mycorrhization tend to increase plant tolerance to other stress (biotic and abiotic) conditions [12].

Despite the major ecological importance of AMF, there is a lack of knowledge on the effect of AMF on soil microbial properties and biological activities, or the effect of ecosystem management (including grazing) on mycorrhization. The complex herbivore–plant–AMF is dependent on the species involved in these interactions, the intensity and frequency of grazing, vegetation, and topography [13,14]. Some studies indicate that grazing influences soil biochemical activity, usually through the degradation of soil structure by trampling [15] and the imbalance of soil chemical, microbial and biochemical properties [16].

There is limited information about the effects of grazing on AMF communities and on soil microbial composition and function, especially for low fertility soils, as found in most Mediterranean arid and semi-arid regions. This work addressed both questions in three arid and semi-arid regions of Tunisia, by evaluation of grazing intensity on AMF–plant interactions, and the consequent impact on soil microbial communities.

2. Materials and Methods

2.1. Study Site

This study was developed at three conserved natural areas (Figure 1) in the arid and semi-arid ecosystem of Tunisia, under Mediterranean climate: Bou-Hedma National Park (Sidi Bouzid coordinates, 34°39′ N, 94°8′ E), Zarat-Gabes protected area (Gabes coordinates, 33°41′ N, 10°23′ E) and Oued Dkouk Natural Reserve (Tataouine coordinates, 32°37′ N, 10°18′ E). The annual temperature range is very high in these areas, with minimal and maximal monthly temperature means of 17 °C (January) and 36 °C (August), respectively. The mean annual rainfall in these ecosystems varies between 100 and 260 mm.

Figure 1. Location of the three conserved natural areas in the arid Mediterranean ecosystem of Tunisia.

Experiments were carried out in three natural protected areas of Tunisia (Bou-Hedma, Zarat, and Oued Dkouk) (Table 1). At each natural park samples were taken in two sites, one inside and the other outside the protected area. The studied sites differed in grazing intensity and soil type (Table 1). The three sites inside the conserved areas were subjected to a light grazing, while the other three sites outside were subjected to more intensive grazing by domestic herds of sheep, goats and camels.

Table 1. Details of the six sampling sites at the three conserved natural areas.

Conserved Areas		Site	Vegetation Type	Grazing Intensity
Bou-Hedma	inside	1	*Acacia* spp. and spontaneous *herbaceous plants*	Light
	outside	2	Cultivated and spontaneous herbaceous plants	Intensive
Zarat	inside	3	*Acacia* spp. and spontaneous herbaceous plants	Light
	outside	4	Cultivated and spontaneous herbaceous plants	Intensive
Oued Dkouk	inside	5	Spontaneous herbaceous plants	Light
	outside	6	Spontaneous herbaceous plants	Intensive

2.2. Roots and Soil Sampling

Five plant species (common to all study sites) and three plants per species per site were analyzed for AMF colonization. The herbaceous plants collected were: *Lotus creticus* (Fabaceae), *Medicago truncatula* (Fabaceae), *Astragalus corrugatus* (Fabaceae), *Malva aegyptiaca* (Malvaceae), *Diplotaxis simplex* (Cruciferacea).

Plant roots were carefully collected in order to access the fine active roots where mycorrhiza colonization occurs. Simultaneously, rhizosphere soil of each plant was also collected. For each site, a pooled soil sample (composed of 5 soil cores with 10 cm diameter and 20 cm length) was collected in an area without vegetation and used as a control (bulk soil) for the influence of the plant in the dynamic soil characteristics. Soils were sieved (2 mm) to remove the remains of plants, gravel and earthworms, and stored at 4 °C for further analysis. Soil samples were analyzed in triplicates.

2.3. AMF Colonization Status

According to the methods of Phillips and Hayman [17], AMF colonization was evaluated after observation of 30 root fragments per plant. Of each plant species, root segments (1–2 cm length) were submerged in 10% potassium hydroxide (KOH) at 90 °C for 45 min. After bleaching and acidification steps with hydrogen peroxide (H_2O_2) and hydrogen chloride (HCl), respectively, root pieces were colored with 1% Trypan Blue solution. The duration of each step varied among plant species, according to the respective root diameter and surface root characteristics. For each plant species from each site, 30 root fragments were placed on slides and preserved with lactoglycerol. In all the stained roots, the presence of hyphae, vesicles, and arbuscules inside the root were viewed through a microscope (Nikon, Tokyo, Japan) at 400× magnification. The frequency (F%): number of colonized roots/total number of observed roots) and intensity of mycorrhization (M%): proportion, in percentage, of the root colonized by AMF) were calculated using the Mycocalc program (http://www2.dijon.inra.fr/mychintec/Mycocalcprg/download.html)

2.4. Quantification of AMF Spore Density

Using the wet sieving method from Gerdemann and Nicolson [18], AMF spores occurring in soil samples were extracted and quantified. Three nested sieves of 1000, 100, and 32 μm were used. For each soil sample, quantities of 100 g were submerged in 1 L of tap water. After a stirring step, the supernatant was sieved through the nested sieves. The spores that hold on to the two sieves of 100 and 32 μm were recovered in 5 mL centrifuge tubes. After a step of centrifugation on a viscosity gradient (sucrose solution at 60%), the supernatant retained was rinsed with distilled water to remove the sucrose solution. Retrieved AMF spores of each soil sample were counted under a stereomicroscope (40× magnification) and average numbers were calculated per 100 g of dry soil.

2.5. Soil Analysis: Physical and Chemical Properties

Soil pH and electrical conductivity were determined in a 1:1 (*v/w*) water: bulk soil suspension using a pH meter (Matest, Treviolo, Italy) and a conductivity meter (Bibby Scientific, Bibby Scientific, Staffordshire, UK), respectively [19]. Soil texture, which represents the granulometric distribution of its constituents (the proportion between small particles: clay, silt and sand), was calculated using the Robinson's pipette method [20]. The other main physicochemical soil characteristics (organic matter, total nitrogen, total phosphorus and water content) were determined by conventional analyses performed by the Soil Analysis Laboratory in the Regional Commissariat for Agricultural Development, in Gabes. The determination of the organic matter was carried out indirectly, starting from the determination of the organic carbon content of soil. The determination of total phosphorus was evaluated by a degradation acid reaction step followed by a dosing step carried out in an automated spectrophotometer (Shimadzu, Kyoto, Japan). The total nitrogen was determined following the Kjeldahl method.

2.6. Microbiological and Biochemical Properties

For the impact of AMF on microbiological parameters, the carbon of the microbial biomass (Cmic) present in the plant rhizosphere and bulk soil were evaluated following the "fumigation–extraction" technique [21]. This method is based on three essential steps: a fumigation step with chloroform, incubation in a 10-days fumigation period, and an extraction step with ninhydrin-N reactive and potassium chloride (KCl). This technique has been used to provide rapid and accurate measurements of soil biomass -C and -N.

To evaluate the effect of AMF on biochemical properties, alkaline phosphatase and β-glucosidase activities were calculated and evaluated according to the method of Caravaca et al. [22] in a spectrophotometer (Shimadzu, Japan) at 398 nm. The dehydrogenase activity was determined as described by Garcia et al. [23] in a spectrophotometer (Shimadzu, Japan) at 490 nm.

2.7. Statistical Analyses

Analyses of variance (ANOVA) for repeated measures using the XLSTAT (v2010.5.04) software (Addinsoft, New York, NY, USA) were ascertained to test the effect and the significant difference between the studied parameters. Least significant difference values at the 5% levels of significance ($p \leq 0.05$) were calculated to assess differences between different values.

To evaluate the effects of the grazing parameter (explanatory variable) on mycorrhizal properties (F%, M% and number of spores) (response variables), ANOVA and Canonical Correlation Analysis (CCorA) were applied.

To evaluate the relationships between soil physical and chemical parameters and mycorrhizal properties (F%, M% and number of spores), a principal-component analysis (PCA) was applied.

Pearson's correlation was used to determine relationships between variables: mycorrhizal properties and microbiological and biochemical parameters.

To model the relationships between mycorrhizal properties and parameters of soil microbial and biochemical activities by linear regression, XLSTAT (v2010.5.04) software was used.

3. Results

3.1. Physical and Chemical Properties of Soils

The more resilient soil characteristics such as soil texture (Table 2) and pH did not show big discrepancies among sites inside and outside the protected areas. However, differences were obtained for the more responsive soil dynamic characteristics: electrical conductivity (E.c), total nitrogen (T.N), total phosphorus (T.P), organic matter (Org. Mat), and water content (Wat. Con). In general, the studied sites had an alkaline pH ranging from 8.0 to 8.4. The highest percentages of soil organic matter were observed in Site 1 (2.6%) as well as inside the other protected areas. Identical patterns were observed for the total nitrogen, the highest level (194 ppm) also being observed in Site 1, followed by Sites 3 and 5 (inside the protected areas). For the total phosphorus content, it was the contrary: the highest values were registered in the sites outside the protected areas (15.3 ppm was the highest value, registered in Site 6). Water content (varying between 1.3 to 3.3%) and electrical conductivity (ranging from 1.3 to 2.5 $s \cdot m^{-1}$) always presented the highest values in the sites inside each protected area.

Table 2. Physical and chemical properties of the six sampling sites from the three conserved ecosystems.

Parameters	Bou-Hedma		Zarat		Oued Dkouk	
	Inside	Outside	Inside	Outside	Inside	Outside
	Site 1	Site 2	Site 3	Site 4	Site 5	Site 6
Clay (%)	11.0 ± 0.2^{b}	12.8 ± 0.1^{a}	9.2 ± 0.2^{d}	10.2 ± 0.1^{c}	6.5 ± 0.2^{f}	7.0 ± 0.1^{e}
Silt (%)	23.5 ± 1.1^{d}	26.1 ± 0.1^{b}	25.1 ± 0.2^{c}	34.7 ± 0.3^{a}	16.4 ± 0.7^{e}	15.3 ± 0.5^{f}
Sand (%)	65.8 ± 0.2^{c}	60.0 ± 1.2^{e}	64.7 ± 1.1^{d}	55.0 ± 1.4^{f}	77.0 ± 1.1^{b}	78.6 ± 1.1^{a}
pH	8.1 ± 0.1^{b}	8.1 ± 0.1^{b}	8.0 ± 0.1^{b}	8.1 ± 0.1^{b}	8.4 ± 0.1^{a}	8.4 ± 0.1^{a}
E.c ($s \cdot m^{-1}$)	2.5 ± 0.3^{a}	1.8 ± 0.2^{c}	2.4 ± 0.3^{a}	1.6 ± 0.2^{d}	2.0 ± 0.1^{b}	1.3 ± 0.1^{e}
T.N (ppm)	191.0 ± 23^{a}	150.0 ± 10^{d}	174.0 ± 15^{b}	130.0 ± 10^{e}	160.0 ± 10^{c}	90.0 ± 5^{f}
T.P (ppm)	7.0 ± 0.1^{f}	10.6 ± 0.2^{c}	8.2 ± 0.2^{e}	12.4 ± 0.2^{b}	10.0 ± 0.2^{d}	15.3 ± 0.6^{a}
Org. Mat (%)	2.6 ± 0.2^{a}	1.4 ± 0.1^{d}	2.0 ± 0.1^{b}	1.2 ± 0.1^{e}	1.6 ± 0.3^{c}	0.8 ± 0.1^{f}
Wa. Cont (%)	3.3 ± 0.1^{a}	2.3 ± 0.1^{d}	2.9 ± 0.2^{b}	2.0 ± 0.1^{e}	2.5 ± 0.1^{c}	1.4 ± 0.3^{f}

E.c: electrical conductivity; T.N: total nitrogen; T.P: total phosphorus; Org. Mat: organic matter; Wat. Con: water content. Letters a–f: significant differences ($p < 0.05$); mean and standard error values ($n = 3$).

3.2. AMF Colonization of Plant Roots

Three plants and 30 root fragments of each one of the five plant species were analyzed per site. Direct observation of the root samples under the microscope showed that the roots of all studied plants, except those of *Diplotaxis simplex*, were colonized by AMF. All the structures characteristic of root colonization by AMF (intracellular aseptate hyphae, vesicles and arbuscules) were observed. The highest mycorrhizal frequency (F%)) was observed for *M. truncatula* in all the studied sites. At each site, the mycorrhizal frequency (F%)) varied among the plant species (Figure 2). The plants with the highest AMF root colonization belong to the Fabaceae family (*M. truncatula*, *A. corrugatus* and *L. creticus*) in all the studied sites, always with higher values in the site inside each conserved area.

There were significant differences ($p < 0.001$) in the mycorrhizal intensity (M%) among sites, again with highest M% for *M. truncatula* as well as, for each plant species, for the plants in the sites inside the conserved areas (Figure 2).

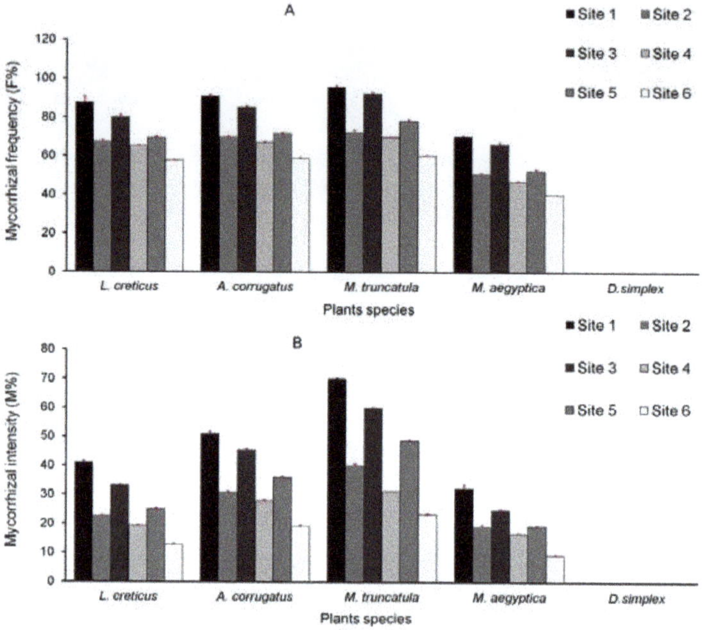

Figure 2. Mycorrhizal frequency (**A**) and Mycorrhizal intensity (**B**) of the five herbaceous plants in the different studied sites. Data are reported as mean (±SE) of three replicates per sample.

3.3. Densities of Spore Populations in the Studied Soils

The density of AMF spores isolated from the rhizosphere of the sampled plants varied between 856 (*M. truncatula*) in Site 1 and 81 spores/100 g of soil (*D. simplex*) in Site 6 (Figure 3). Therefore, spore density varied significantly ($p < 0.001$) among the studied sites; the maximum values were recorded in the rhizosphere of plants from sites 1, 3 and 5 (inside the conserved areas and lightly grazed) and the minimum values in sites 2, 4 and 6 (outside the conserved areas and intensively grazed). Bare areas (bulk soil), followed by *D. simplex* rhizosphere soil, always presented the lowest values for the six sites (Figure 3).

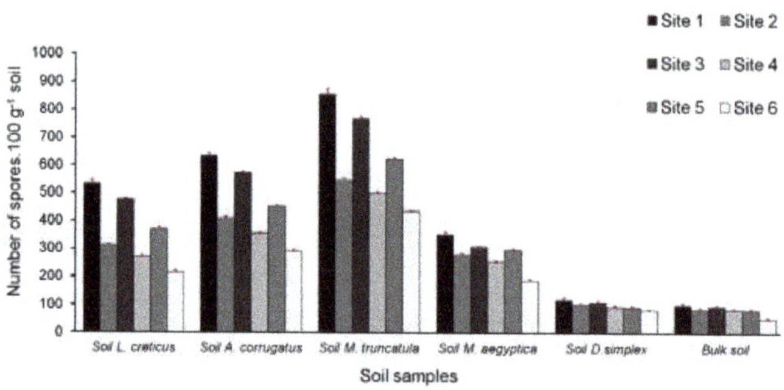

Figure 3. Distribution of AMF spores in the different rhizosphere soils. Data are reported as mean (±SE) of three replicates per sample.

There was a clear relationship between the intensity of root cortex AMF colonization (M%), mycorrhizal frequency (F%), and the density of AMF spores in the rhizospheres (Figures 2 and 3). Furthermore, the highest spore density was registered in the rhizosphere of the plant species with higher mycorrhiza frequency and intensity (*M. truncatula*).

3.4. Effect of the Grazing Intensity on the Different AMF Parameters

The grazing intensity strongly affected the mycorrhizal colonization and density of spores (Table 3), in accordance with the lowest values always registered, either for the plant mycorrhizal status or for AMF spore populations, in samples from sites outside the protected areas (Figures 2 and 3). This explains the negative effect of the grazing intensity on the different mycorrhizal parameters (Figure 4).

Table 3. Two-factor ANOVA analysis of the impact of grazing intensity on mycorrhizal parameters.

Factor	Grazing Intensity	
	F-Ratio	*p*-Value
Mycorrhizal frequency (F%)	3.52	0.07 ns
Mycorrhizal intensity (M%)	11.26	0.001 ***
AMF spore density	10.55	0.002 **

*** Significant at $p < 0.001$, ** Significant at $p < 0.01$, ns: non-significant.

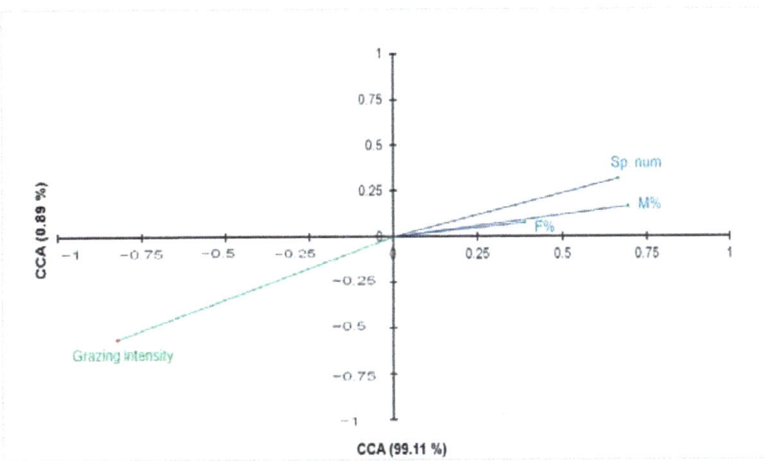

Figure 4. Results of a Canonical Correlation Analysis (CCorA) for the relationships between grazing intensity and AMF properties. Sp. num: number of spores; F%: mycorrhizal frequency; M%: mycorrhizal intensity.

3.5. Effect of Soil Properties on the Different AMF Parameters

According to the results of the Principal-component analysis (PCA) (Figure 5), the highest values of AMF colonization and number of spores were found in Site 1 at Bou-Hedma followed by Site 3 at Zarat. The major physical and chemical parameters of these sites (Org. Mat, E.c, Wat. Con, T.N) were in a strongly positive correlation with the different mycorrhizal properties. In contrast, the total phosphorus (T.P) available negatively affected all the mycorrhizal parameters and was considered a limited factor to these parameters (Figure 5).

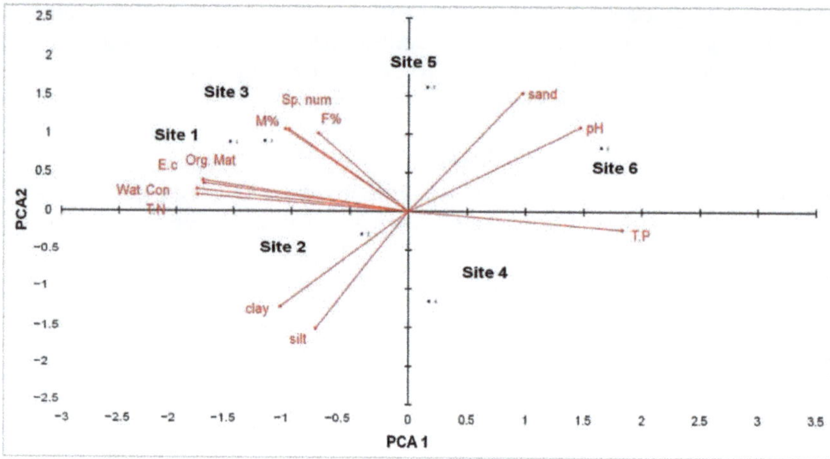

Figure 5. Results of a principal-component analysis (PCA) showing the relationships between soil properties and AMF parameters. E.c: electrical conductivity; T.N: total nitrogen; T.P: total phosphorus; Org. Mat: organic matter; Wat. Con: water content; Sp. num: number of spores; F%: mycorrhizal frequency; M%: mycorrhizal intensity.

3.6. *Microbiological and Biochemical Properties of the Soils and Impact of AMF on Soil Microbial Communities*

We found significant effects ($p < 0.001$) of the mycorrhizal plants on Cmic (Figure 6), this meaning that plants are affecting soil microbial biomass and consequently potential soil dynamics. The Cmic values were lower in bare areas (control soil) than in rhizosphere soils, the maximum values being recorded in the rhizospheres of the three Fabaceae plants *M. truncatula, A. corrugatus* and *L. creticus*. Cmic also markedly decreased in the intensively grazed sites (outside the protected areas), following the same trend of mycorrhizal frequency and intensity in root samples and of number of spores in rhizosphere soils (Figures 2, 3, 6 and 7).

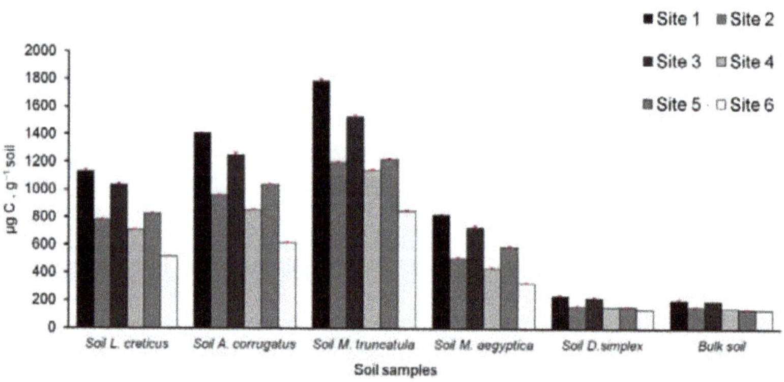

Figure 6. Microbial biomass carbon (Cmic) in the different sampled soils. Data are reported as mean (±SE) of three replicates per sample.

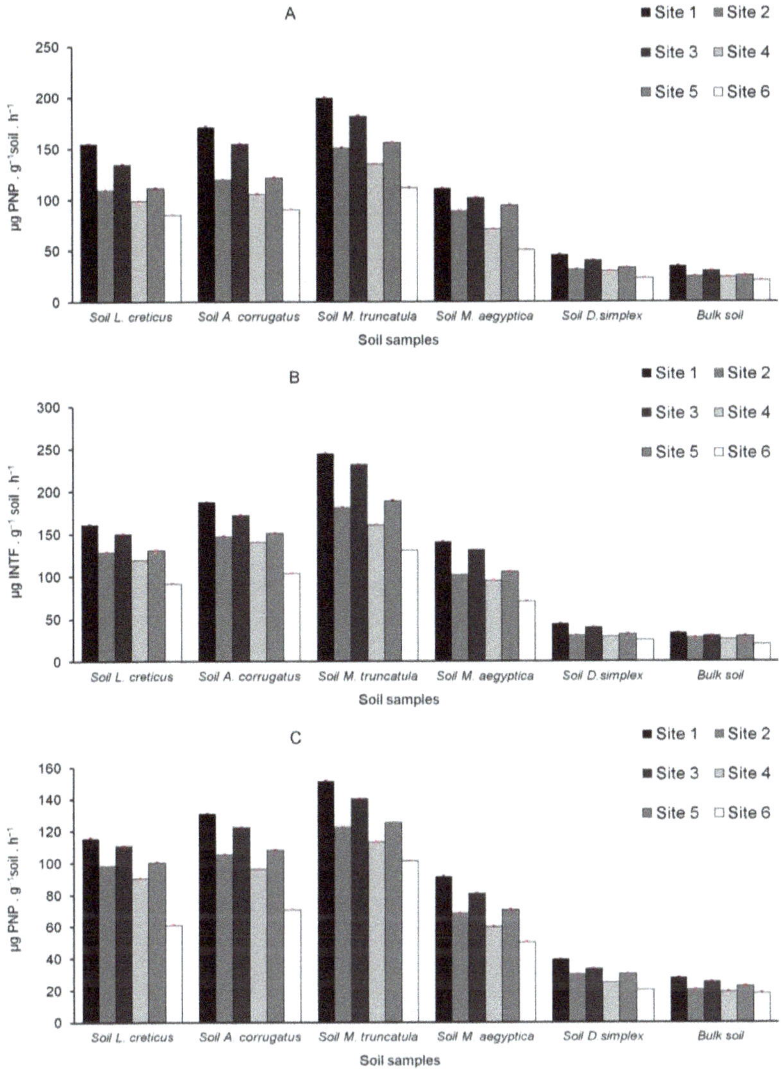

Figure 7. Phosphatase (**A**), dehydrogenase (**B**), β-glucosidase (**C**) activities in the studied soils. PNP: p nitrophenol, INTF: iodonitrotetrazolium formazan. Data are reported as mean (±SE) of three replicates per sample.

The highest activities of the three enzymes evaluated (phosphatase, dehydrogenase and β-glucosidase) (Figure 7) were observed in the rhizospheres of the three Fabaceae plants *M. truncatula*, *A. corrugatus* and *L. halophilus*, following the same trend of Cmic for each plant species and each site (Figure 6) as well as the values of mycorrhizal frequency and intensity and number of spores (Figures 2, 3 and 7), evidencing activities consistently greater in lightly grazed sites in comparison with those that were intensively grazed (Figures 2, 3, 7 and 8). The ANOVA analysis evidenced significant differences in these values.

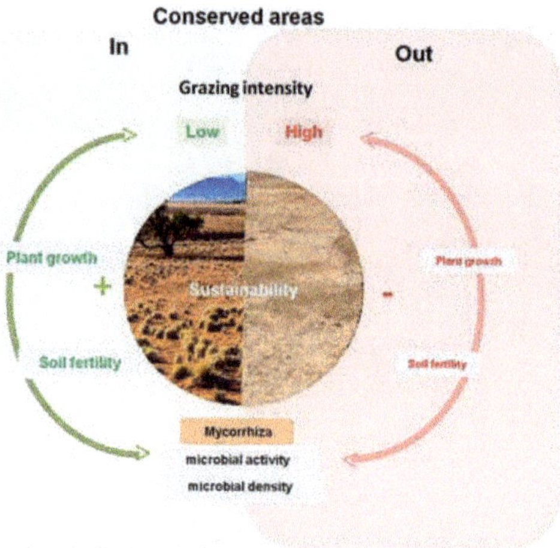

Figure 8. Schematic representation of the AMF effects on soil fertility and ecosystem stability under lightly grazed sites (inside the protected areas) versus intensively grazed sites (outside the protected areas), for arid Mediterranean ecosystems.

A clear relationship was observed between the AMF parameters (frequency and intensity of colonization of roots and number of AMF spores in the rhizosphere) and microbial activity (microbial biomass and enzyme activities) (Table 4, Figures S1 and S2). Based on the Pearson correlation coefficient, positive correlations with a highly significant value ($p < 0.001$) were observed between all AMF parameters (F%, M% and number of spores) and microbial activity (microbial biomass and enzyme activities: dehydrogenase, phosphatase and the β-glucosidase). A lower significant value ($p < 0.05$) was only observed in Site 6 (outside Oued Dkouk Natural Reserve) between the frequency of AMF colonization and two microbial parameters: microbial biomass and activity of β-glucosidase (Table 4).

Table 4. Pearson correlation coefficient between frequency of mycorrhization (F%), intensity of mycorrhization (M%) and density of AMF spores (Sp. num) and microbial activity expressed by microbial biomass (Cmic), phosphatase activity, dehydrogenase activity, and β-glucosidase activity.

Conserved Areas		Title	Cmic	Phosphatase	Deydrogenase	β-Glucosidade
Bou Hedma	Site 1	F%	0.93 ***	0.97 ***	0.95 ***	0.96 ***
		M%	0.99 ***	0.99 ***	0.99 ***	0.98 ***
		Sp. num	0.93 ***	0.97 ***	0.95 ***	0.96 ***
	Site 2	F%	0.93 ***	0.97 ***	0.96 ***	0.96 ***
		M%	0.98 ***	0.98 ***	0.99 ***	0.99 ***
		Sp. num	0.98 ***	0.96 ***	0.98 ***	0.98 ***
Zarat	Site 3	F%	0.95 ***	0.97 ***	0.96 ***	0.97 ***
		M%	0.99 ***	0.98 ***	0.99 ***	0.98 ***
		Sp. num	0.95 ***	0.97 ***	0.96 ***	0.97 ***
	Site 4	F%	0.92 ***	0.97 ***	0.97 ***	0.96 ***
		M%	0.96 ***	0.97 ***	0.99 ***	0.98 ***
		Sp. num	0.97 ***	0.96 ***	0.97 ***	0.95 ***
Oued Dkouk	Site 5	F%	0.96 ***	0.97 ***	0.97 ***	0.97 ***
		M%	0.99 ***	0.99 ***	0.98 ***	0.97 ***
		Sp. num	0.98 ***	0.97 ***	0.99 ***	0.98 ***
	Site 6	F%	0.90 *	0.94 ***	0.95 ***	0.91 *
		M%	0.98 ***	0.98 ***	0.99 ***	0.98 ***
		Sp. num	0.98 ***	0.99 ***	0.97 ***	0.95 ***

*** Significant at $p < 0.001$, * Significant at $p < 0.05$.

The results shown above, concerning the effect of grazing on AMF colonization (Table 3, Figure 4) and the impact of AMF on microbial communities and activities (Table 4, Figures S1 and S2), are summarized in Figure 8, which evidences the importance of AMF on soil functionality under arid Mediterranean ecosystems. Significant mycorrhizal colonization of roots was registered in lightly grazed sites, leading to high levels of microbial communities expressed by high values of biochemical activities. All these necessarily enhances soil fertility, better-adaptation and growth of plants and finally, in the stability of the ecosystem (Figure 8).

4. Discussion

Mediterranean arid and semi-arid ecosystems are characterized by high temperatures and drought for most of the year. These conditions limit plant establishment and growth and accelerate soil degradation and change microbial communities [1]. Mycorrhiza form communication pathways between plants and soil, influencing plant nutrient cycling, and restoring and maintaining soil fertility, thus influencing the microbial communities of the rhizosphere and extending the influence of plants to the soil. Several works emphasized the role of AMF in sustaining plant cover in semi-arid and arid ecosystems [4,8,10] as is the case of the following Mediterranean conserved areas in Tunisia: Bou-Hedma National Park, Zarat protected area and Oued Dkouk Natural Reserve.

Under natural conditions, about 90% of the terrestrial plants are mycorrhized, and AMF are found in all climates and ecosystems [10]. Due to their role on plant nutrition and defense, and to the importance of the extra-radical mycelium in the establishment of biological networks, AMF are determinant for the establishment and sustainability of plant communities and environment functioning [7]. However, the diversity of the AMF community and the intensity of arbuscular mycorrhizal colonization of natural vegetation is dependent on the availability of AMF spores and the mycorrhizal dependency of the plant species [24,25], as well as on the soil structure and management [26]. All these statements agree with the results of the present study, which evidenced colonization by AMF for all the herbaceous species, except for the Cruciferaceae *Diplotaxis simplex* (Figure 2). This was not surprising since Cruciferaceae are usually recognized as non-mycorrhizal plants [27]. However, distinct AMF colonization rates and intensities were observed for each plant species, which may be related to different levels of mycorrhizal dependence [28], and/or the availability of AMF spores. Higher mycorrhization rates were observed in the legume species *M. truncatula, A. corrugatus* and *L. creticus*, which presented high levels of mycorrhizal intensity, independent of the sampled site (Figure 2). This may highlight their mycorrhizal dependency and high demand for phosphorus (P) in comparison with plants from other families such as Poaceae [29].

In general, and particularly in semi-arid and arid ecosystem, AMF vary greatly with soil characteristics. Several biotic and abiotic factors may contribute to the distinct mycorrhizal intensities observed for the same plant species [30,31]. One abiotic factor that severely interferes with mycorrhization is the concentration of available phosphorus [10,32,33]. In the present study, soil phosphorus concentrations varied between sites, ranging from 7 to 15.3 ppm, and could explain the distinct mycorrhizal intensities observed for the same species in distinct sites (Figures 2 and 5), supporting that its availability is a crucial driver of mycorrhizal communities and activities [34].

Another factor frequently cited as responsible for lower levels of mycorrhiza formation is the availability of AMF spores, which is known to greatly vary in the ecosystems [35]. Sporulation in AMF occurs when the development of the mycelium begins to be limited by nutrients and is a highly carbon demanding process. This may explain why the number of AMF spores in the rhizosphere varies among plant species and, for the same plant species, among sites [36–38]. Apart from the rate of spore formation, the number of AMF spores in a soil also depends on the rates of spore germination and degradation [39]. As AMF are obligate biotrophs, the number of spores and propagules tends to be higher in the rhizosphere than in the bulk soil [12,40], and higher in the rhizosphere of plants with a higher intensity of AMF colonization (Figure 3). These differences are particularly evident in arid and semi-arid soils with high organic matter turnover rates and low organic matter content [41].

In low fertility soils, mycorrhiza are highways for nutrient and water transport, expanding the plant root system and the volume of soil exploited by the plant. However, similarly to roots, hyphae are leaky and lose nutrients into the hyphosphere, which will promote the development of selected microorganisms. In this context, it is expected that AMF play a crucial role in the biological characteristics of the rhizosphere [22,42] (Figure 6). The high values of Cmic observed in the rhizosphere soil of mycorrhizal plants imply that mycorrhiza contribute to improve the availability of carbon substrates to the microbial community of the rhizosphere [43,44]. These results support the hypothesis that AMF establish unique interactions with plant roots, conferring special characteristics to the rhizosphere [45], where several by product-based symbiosis and microbial loops may be assembled, contributing to improved carbon use efficiency. In this particular respect, the importance of AMF in promoting the development of the microbial community is confirmed by the low and comparable levels of microbial carbon observed in the soils without plant cover (control) and in the rhizosphere soil of the non-mycorrhizal plant *D. simplex* (Figure 6).

AMF increase the diversity of the carbon sources available to the microorganisms in the rhizosphere [46]—which is partly due to their nutritional mode; the excretion of catabolic enzymes to the surrounding medium, and to the direct access by the AMF to the plant carbon. Phosphatase and β-glucosidase are two of those catabolic enzymes, and their activities were in fact higher in the rhizosphere of the mycorrhizal plants than in that of the non-mycorrhizal plant studied or in the bulk soil (Figure 5). The importance of the soil microbial activity in association with several enzymes' activities was highlighted by the similarity in the activity patterns of the two hydrolytic enzymes (phosphatase and β-glucosidase) as well as of dehydrogenase, indicators of microbial activity [47], which can be inferred as being decreased under high grazing intensity (Figures S1 and S2).

At this point, it is clear that plant and AMF species are important modulators of rhizosphere characteristics (Figures 6–8). What is not clear is why biological indicators of soil characteristics are so distinct in the three sites outside the conserved areas. One key factor common to these sites may be limiting and related to mycorrhizal colonization. Grazing may affect mycorrhization (Table 3, Figure 4) directly in nutrient-limited ecosystems through a direct competition for carbon (Figure 8). Grazing was consistently associated with lower soil organic matter and increased P concentrations showing a strong impact on soil chemical properties, able to negatively influence mycorrhization in a direct way through the increase of P bioavailability. But in nutrient-limited ecosystems grazing per se may also affect mycorrhization (Table 3, Figure 4).

Plants are the primary producers of ecosystems, obtaining their biomass and energy from the carbon fixed by photosynthesis (Figure 8). Depending on the plant species and on the growth conditions, 20–50% of the newly photosynthetically fixed carbon may be transported to the roots and lost as rhizodeposition. In arid soils, this carbon is the main source of energy and biomass building material for the microbial community. Therefore, increasing grazing intensity decreases the carbon availability from plants and, consequently, also decreases the carbon available for mycorrhization [13,14]. An increase in grazing intensity has impacts on soil biological characteristics, with a significant decrease in the mycorrhizal intensity, as evidenced in the present study, for the three mycorrhizal plant species sampled in the sites outside the conserved areas (Table 3, Figure 5; Figure 8).

5. Conclusions

AMF is of vital importance in the arid Tunisian ecosystem, where soils are generally poor in organic matter and nutrients. High mycorrhization levels and high spore densities were revealed across the different sites, with particular incidence in the sites inside the conserved areas. Therefore, grazing intensity affected the potential beneficial influence of AMF on soil microbial processes. At this point, the importance of AMF to ecosystem dynamics is clear. What is not clear is how AMF are adapted to grazing intensity to allow ecosystem stability, particularly for situations that may be represented by the three sites outside the conserved areas. Therefore, future research should be focused on determining the ways through which AMF adapt to grazing intensity. This may be a key feature of ecosystem management, considering that AMF could be an important alternative for sustaining soil quality, and could be exploited as potential inoculants for rehabilitation and restoration programs in Mediterranean ecosystems.

Supplementary Materials: The following are available online at http://www.mdpi.com/1424-2818/12/6/234/s1, Figure S1: Linear regression to model the correlations between mycorrhizal proprieties (mycorrhizal frequency, mycorrhizal intensity, and number of spores) and the microbiological parameter microbial biomass carbon. The analyses included all available data: from the rhizospheres of the four plants and from bulk soil, Figure S2: Linear regression to model the correlations between mycorrhizal proprieties (mycorrhizal frequency, mycorrhizal intensity, and number of spores) and biochemical activities: phosphatase (in µg PNP. g^{-1} soil·h^{-1}), dehydrogenase (µg INTF. g^{-1} soil·h^{-1}) and β-glucosidase (µg PNP. g^{-1} soil·h^{-1}). The analyses included all available data: from the rhizospheres of the four plants and from bulk soil.

Author Contributions: Conceptualization, M.M. (Mohamed Mars); methodology, N.M.; validation, M.M. (Mosbah Mahdhi) and T.D.; formal analysis, N.M.; investigation, N.M.; data curation, N.M.; writing—original draft preparation, N.M.; writing—review and editing, C.C. and M.F.C.; visualization, C.C. and M.F.C.; project administration, M.M. (Mohamed Mars); funding acquisition, C.C. and M.F.C. All authors have read and agreed to the published version of the manuscript.

Funding: This work was supported by the Ministry of High Education and Research Development-Tunisia and by FCT/MCTES for the financial support to cE3c (project UIDB/00329/2020) and CESAM (projects UIDP/50017/2020+UIDB/50017/2020), through Portuguese national funds.

Acknowledgments: We thank the personal staff of Bou-Hedma National Park, of Zarat-Gabes Protected area and of the Natural reserve of Oued Dkouk, for facilities and support for root and soil sampling. Thanks, are also due to Francisco Caeiro for providing helpful comments and English revision.

Conflicts of Interest: The authors declare no conflict of interest.

References

1. Martínez-García, L.; Miranda, J.; Pugnaire, F.I. Impacts of changing rainfall patterns on mycorrhizal status of a shrub from arid environments. *Eur. J. Soil Biol.* **2011**, *50*, 64–67. [CrossRef]
2. Mahmoudi, N.; Cruz, C.; Mahdhi, M.; Mars, M.; Caeiro, M.F. Arbuscular mycorrhizal fungi in soil, roots and rhizosphere of Medicago truncatula: Diversity and heterogeneity under semi-arid conditions. *PeerJ* **2019**. [CrossRef] [PubMed]
3. Mosbah, M.; Philippe, D.L.; Mohamed, M. Molecular identification of arbuscular mycorrhizal fungal spores associated to the rhizosphere of Retama raetam in Tunisia. *Soil Sci. Plant. Nutr.* **2018**, *64*, 335–341. [CrossRef]
4. Brundrett, M. Mycorrhizal associations and other means of nutrition of vascular plants: Understanding the global diversity of host plants by resolving conflicting information and developing reliable means of diagnosis. *Plant Soil* **2009**, *320*, 37–77. [CrossRef]
5. Van der Heijden, M.G.A.; Martin, F.M.; Selosse, M.A.; Sanders, J.R. Mycorrhizal ecology and evolution: The past, the present, and the future. *New Phytologist.* **2015**, *204*, 1406–1423. [CrossRef] [PubMed]
6. Barea, J.M.; Azcón, R.; Azcón-Aguilar, C. Mycorrhizosphere interactions to improve plant fitness and soil quality. Antonie van Leeuwenhoek. *Inter. J. Gen. Mol. Microb.* **2002**, *81*, 343–351.
7. Entry, J.A.; Rygiewicz, P.T.; Watrud, L.S.; Donnelly, P.K. Influence of adverse soil conditions on the formation and function of arbuscular mycorrhizas. *Adv. Environ. Res.* **2002**, *7*, 123–138. [CrossRef]
8. Smith, S.E.; Read, D.J. *Mycorrhizal Symbiosis*; Academic Press: San Diego, CA, USA, 1997.
9. Van Aarle, I.M.; Cavagnaro, T.R.; Smith, S.E.; Smith, F.A.; Dickson, S. Metabolic activity of Glomus intraradices in Arumand Paris-type arbuscular mycorrhizal colonization. *New Phytologist.* **2005**, *166*, 611–618. [CrossRef] [PubMed]
10. Smith, S.E.; Read, D.J. *Mycorrhizal Symbiosis*; Academic Press: San Diego, CA, USA, 2008.
11. Bago, B.; Vierheilig, H.; Piché, Y.; Ázcon-Aguilar, C. Nitrate depletion and pH changes induced by the extraradical mycelium of the arbuscular mycorrhizal fungus Glomus intraradices grown in monoxenic culture. *New Phytol.* **1996**, *133*, 273–280. [CrossRef] [PubMed]
12. Azcón-Aguilar, C.; Palenzuela, E.J.; Roldan, A.; Bautista, S.; Vallejo, R.; Barea, J.M. Analysis of the mycorrhizal potential in the rhizosphere of representative plant species from desertification-threatened Mediterranean shrub lands. *Appl. Soil Ecol.* **2003**, *22*, 29–37. [CrossRef]

13. Barto, E.K.; Rilling, M.C. Does herbivory really suppress mycorrhiza? A meta-analysis. *J. Ecol.* **2010**, *98*, 745–753. [CrossRef]
14. Mendoza, R.; Cabello, M.; Anchorena, J.; García, M.L. Soil parameters and host plants associated with arbuscular mycorrhizae in the grazed Magellanic steppe of Tierra del Fuego. *Agric. Ecosyst. Environ.* **2011**, *140*, 411–418. [CrossRef]
15. Raiesi, F.; Asadi, E. Soil microbial activity and litter turnover in native grazed and ungrazed rangelands in a semiarid ecosystem. *Biol. Fertil. Soil* **2006**, *43*, 76–82. [CrossRef]
16. Liao, J.D.; Boutton, T.W. Soil microbial biomass response to woody plant invasion of grassland. *Soil Biol. Biochem.* **2008**, *40*, 1207–1216. [CrossRef]
17. Phillips, J.M.; Hayman, D.S. Improved procedure for clearing roots and staining parasitic and vesicular arbuscular mycorrhizal fungi for rapid assessment of infection. *Trans. Br. Mycol. Soc.* **1970**, *55*, 158–161. [CrossRef]
18. Gerdemann, J.W.; Nicolson, T.H. Spores of mycorrhizal Endogone species extracted by wet sieving and decanting. *Trans. Br. Mycol. Soc.* **1963**, *46*, 235–244. [CrossRef]
19. AFNOR. *Recueil de normes françaises, qualité des sols, méthodes d'analyses*, 1st ed.; Association française de normalisation (Afnor): Paris, France, 1987.
20. Naanaa, W.; Susini, J. *Méthodes d'analyse physique et chimique des sols. ES 252, Direction des Sols*; Ministère de l'Agriculture: Tunis, Tunisie, 1988.
21. Amato, M.; Ladd, J.N. Assay for microbial biomass based on ninhydrin-reactive nitrogen in extracts of fumigated soils. *Soil Biol. Biochem.* **1988**, *20*, 107–114. [CrossRef]
22. Caravaca, F.; Alguacil, M.; Torres, P.; Roldán, A. Plant type mediates rhizospheric microbial activities and soil aggregation in a semiarid Mediterranean salt marsh. *Geoderma* **2005**, *12*, 338–375. [CrossRef]
23. Garcia, C.; Hernandez, T.; Costa, F. Potential use of dehydrogenase activity as an index of microbial activity in degraded soils. Commun. *Soil Sci. Plan.* **1997**, *28*, 123–134. [CrossRef]
24. Zhu, Y.G.; Laidlaw, A.S.; Christie, P.; Hammond, M.E. The specificity of arbuscular mycorrhizal fungi in perennial ryegrass, white clover pasture. *Agric. Ecosyst. Environ.* **2000**, *77*, 211–278. [CrossRef]
25. Duponnois, R.; Founoune, H.; Masse, D.; Pontanier, R. Inoculation of *Acacia holosericea* with ectomycorrhizal fungi in a semi-arid site in Senegal: Growth response and influences on the mycorrhizal soil infectivity after 2 years plantation. *For. Ecol. Manag.* **2005**, *207*, 351–362. [CrossRef]
26. Lekberg, Y.; Koide, R.T.; Rohr, J.R.; Aldrich-Wolfe, L.; Morton, J.B. Role of niche restrictions and dispersal in the composition of arbuscular mycorrhizal fungal communities. *J. Ecol.* **2007**, *95*, 95–105. [CrossRef]
27. Muchovej, R.M. *Importance of Mycorrhizae for Agricultural Crops, SS-AGR-170*; Agronomy Department, Florida Cooperative Extension Service; Institute of Food and Agricultural Sciences, University of Florida: Gainesville, FL, USA, 2004.
28. Strullu, D.G. *Les mycorhizes des arbres et des plantes cultivées*; Techniques et Documentation Lavoisier: Paris, France, 1991.
29. Bagayogo, M. *Site Specific Effects of Cereals/Legume Rotations in West Africa: Soil Mineral Nitrogen, Mycorrhizae and Nematodes*; Verlag Graner: Stuttgard, Germany, 1999.
30. Henriques, R.P.; Hay, J.D. The plant communities of a foredune in southeastern Brazil. *Can. J. Bot.* **1998**, *76*, 1323–1330.
31. Mahesh, V.; Selvaraj, T. Occurrence and distribution VA mycorrhizal fungi in the soils polluted with Tannery Effluent. *Adv. Biotech.* **2008**, *8*, 34–36.
32. Dickson, S.; Smith, S.E.; Smith, F.A. Characterization of two arbuscular mycorrhiza fungi in symbiosis with *Allium porum*: Colonization, plant growth and phosphate uptake. *New Phytol.* **1999**, *144*, 163–172. [CrossRef]
33. Bouamri, B.; Dalpé, Y.; Serrhini, M.N.; Bennani, A. Arbuscular Mycorrhizal fungi species associated with rhizosphere of Phoenix dactylifera L. in Morocco. *Afr. J. Boitech.* **2006**, *5*, 510–516.
34. Liu, W.; Zhang, Y.; Jiang, S.; Deng, Y.; Christic, P.; Murray, P.J.; Li, X.; Zhang, J. Arbuscular mycorrhizal fungi in soil and roots respond differently to phosphorus inputs in an intensively managed calcareous agricultural Soil. *Sci. Rep.* **2016**, *6*, 1–11. [CrossRef] [PubMed]
35. Li, L.; Yang, A.; Zhao, Z. Seasonality of arbuscular mycorrhizal symbiosis and dark septate endophytes in grassland site in Southwest China. *Microbial. Ecol.* **2005**, *54*, 367–373.
36. Houngnandan, P.; Yemadje, R.G.; Kane, A.; Boeckx, P.; Van Cleemput, O. Les glomales indigènes de la forêt claire à Isoberliniadoka (Craib et Stapf) à Wari-Maro au centre du Bénin. *Tropicultura* **2009**, *27*, 83–87.

37. Sghir, F.; Chliyeh, M.; Kachkouch, W.; Khouader, M.; OuazzaniTouhami, A.; Benkirane, R.; Douira, A. Mycorrhizal status of *Olea europaea* ssp. Oleaster in Morocco. *J. Appl. Biosci.* **2013**, *61*, 4478–4489. [CrossRef]
38. Al-Areqi, A.N.; Chliyeh, M.; Sghir, F.; Ouazzani, A.; Benkirane, R.T.; Douira, A. Diversity of Arbuscular mycorrhizal fungi in the rhizosphere of Coffea Arabica in the Republic of Yemen. *J. Appl. Biosci.* **2013**, *64*, 4888–4901. [CrossRef]
39. Gemma, J.N.; Koske, R.E.; Carreiro, M. Seasonal variation in spore abundance and dormancy of Gigasporagigantea in Mycorrhizal inoculum potential of a dune Soil. *Mycologia* **1989**, *80*, 211–216. [CrossRef]
40. Lovelock, C.E.; Andersen, K.; Morton, J.B. Arbuscular mycorrhizal communities in tropical forests are affected by host tree species and environment. *Oecologia* **2003**, *135*, 268–297. [CrossRef] [PubMed]
41. Mohammad, M.J.; Hamad, S.R.; Malkawi, H.I. Population of arbuscular mycorrhizal fungi in semi-arid environment of Jordan as influenced by biotic and abiotic factors. *J. Arid Environ.* **2003**, *53*, 409–417. [CrossRef]
42. Cruz, R.E.; Garcia, M.U. Nitrogen fixation and mycorrhizae in acacias on degraded grasslands. In *Tropical Acacias in East Asia and the Pacific*; Kamiset, A., Taylor, D.A., Eds.; Winrock Intenational Institute for Agriculture Research: Bangkok, Thailand, 1992; pp. 59–71.
43. Böhme, L.; Langer, U.; Böhme, F. Microbial biomass, enzyme activities and microbial community structure in two European long-term field experiments. *Agric. Ecosyst. Environ.* **2005**, *109*, 141–152. [CrossRef]
44. Flieβbach, A.; Oberholzer, H.R.; Gunst, L.; Mäder, P. Soil organic matter and biological soil quality indicators after 21 years of organic and conventional farming. *Agric. Ecosyst. Environ.* **2007**, *118*, 273–274. [CrossRef]
45. Dennis, P.G.; Miller, A.J.; Hirsch, P.R. Are root exudates more important than other sources of rhizodeposits in structuring rhizosphere bacterial communities? *Microbial. Ecol.* **2010**, *72*, 313–327. [CrossRef] [PubMed]
46. Rillig, M.C. Arbuscular mycorrhizae, glomalin and soil quality. *Can. J. Soil Sci.* **2004**, *84*, 355–363. [CrossRef]
47. Nannipieri, P.; Kandeler, E.; Ruggiero, P. Enzyme activities and microbiological and biochemical processes in Soil. In *Enzymes in the Environment*; Activity, Ecology and Applications; Burns, R.G., Dick, R.P., Eds.; Marcel Dekker: New York, NY, USA, 2002; pp. 1–33.

diversity

MDPI

Article

Fungal Diversity in the Phyllosphere of *Pinus heldreichii* H. Christ—An Endemic and High-Altitude Pine of the Mediterranean Region

Jelena Lazarević [1,*] and Audrius Menkis [2]

1 Biotechnical faculty, University of Montenegro, Mihaila Lalića 15, 81 000 Podgorica, Montenegro
2 Department of Forest Mycology and Plant Pathology, Uppsala BioCenter, Swedish University of Agricultural Sciences, P.O. Box 7026, SE-75007 Uppsala, Sweden; audrius.menkis@slu.se
* Correspondence: enalazarevic@ac.me

Received: 30 March 2020; Accepted: 21 April 2020; Published: 28 April 2020

Abstract: *Pinus heldreichii* is a high-altitude coniferous tree species naturaly occurring in small and disjuncted populations in the Balkans and southern Italy. The aim of this study was to assess diversity and composition of fungal communities in living needles of *P. heldreichii* specifically focusing on fungal pathogens. Sampling was carried out at six different sites in Montenegro, where 2–4 year-old living needles of *P. heldreichii* were collected. Following DNA isolation, it was amplified using ITS2 rDNA as a marker and subjected to high-throughput sequencing. Sequencing resulted in 31,831 high quality reads, which after assembly were found to represent 375 fungal taxa. The detected fungi were 295 (78.7%) Ascomycota, 79 (21.0%) Basidiomycota and 1 (0.2%) Mortierellomycotina. The most common fungi were *Lophodermium pinastri* (12.5% of all high-quality sequences), *L. conigenum* (10.9%), *Sydowia polyspora* (8.8%), *Cyclaneusma niveum* (5.5%), Unidentified sp. 2814_1 (5.4%) and *Phaeosphaeria punctiformis* (4.4%). The community composition varied among different sites, but in this respect two sites at higher altitudes (harsh growing conditions) were separated from three sites at lower altitudes (milder growing conditions), suggesting that environmental conditions were among major determinants of fungal communities associated with needles of *P. heldreichii*. Trees on one study site were attacked by bark beetles, leading to discolouration and frequent dieback of needles, thereby strongly affecting the fungal community structure. Among all functional groups of fungi, pathogens appeared to be an important component of fungal communities in the phyllosphere of *P. heldreichii*, especially in those trees under strong abiotic and biotic stress.

Keywords: needle pathogens; high altitude forests; DNA metabarcoding; Montenegro

1. Introduction

Pinus heldreichii is a high-altitude (grows at ca. 1200–2000 m) conifer tree species with a discontinuous and restricted distribution in the Mediterranean region. Its forests are naturaly regenerated and consists of small and disjuncted populations located in high mountain areas influenced by the Mediterranean climate in the Balkans and southern Italy. Although in the past *P. heldreichii* formed a continuous forest belt in the Balkans, currently its forests are scattered and largely isolated. Growing primary on shallow calcareous soils, it inhabits typical tree line habitats, often on steep ridges, mountain sides and screens. Such habitats are nutrient poor, exposed, dry and cold during the winter [1,2]. Larger *P. heldreichii* forests can still be found on mountain plateaus or in valleys situated at altitudes of ca. 1200–1300 m that are characterised by more developed soils (cambisols). Those forests are considered to be climazonal. i.e., characterised by permanent and stable vegetation forests of *P. heldreichii*, that are growing at ecological optimum for the species [3]. Being a tertiary relic, *P. heldreichii* evolved to survive severe winter frosts, and short, dry and warm summers with intensive sun radiation. Due to the short growing season, *P. heldreichii* grows very slowly, reaching ca. 15 m height after about 150 years [1]. It develops thick and branched roots, which penetrate deep in cracks of calcareous stones [2]. In the last few centuries, *P. heldreichii* forests were affected by intensive livestock grazing, exploitation and forest fires [4]. Intensive grazing, as well as the short and dry growing season have also resulted in very limited natural regeneration [2–4], though recent observations show some regeneration in the abandoned mountain areas [5].

Pinus heldreichii is a protected species both in Balkan countries and in Italy due to the key importance of nature conservation, protection against gravitational natural hazards, landscape conservation and recreation. Hence, *P. heldreichii* requires special attention, i.e., the development and application of conservation measures [1–4,6].

Fungi represent the largest microbial component associated with forest trees. They play key roles in forest ecosystems, especially in pines forest that are obligatory mutualistic, and are important contributors to the primary production and carbon, nutrient and water cycling [7–9]. Pathogenic fungi may negatively affect health and growth of forest trees [10,11], while fungal endophytes and epiphytes support ecological adaptations of host plants and constitute an important component of microbial biodiversity [12,13]. However, information about fungal communities associated with *P. heldreichii* is limited. Previous studies on needle pathogens of *P. heldreichii* have focused on either those that were affecting natural regeneration [14–16], or potentially invasive ones such as *Dothistroma septosporum* [17]. Information about the occurrence of *D. septosporum* in stands of *P. heldreichii* was not known until recently [17], what indicated the need for a wider assessment of fungal diversity, including fungal pathogens associated with the phyllosphere of this tree species.

The aim of this study was to assess the diversity and composition of fungal communities in living needles of *P. heldreichii* specifically focusing on fungal pathogens. Needles were sampled across the natural distribution range of *P. heldreichii* in Montenegro, including sites under different environmental conditions. This was expected to demonstrate potential site-specific effects of environmental conditions on fungal diversity and community composition. By using high-throughput sequencing of fungal ITS2 rDNA, we examined fungal communities in 2–4 year-old living needles of *P. heldreichii* from six sites situated in four mountain areas in Montenegro.

2. Materials and Methods

2.1. Study Sites

Samples were taken at six sites that were situated in four mountain areas, representing the *P. heldreichii* distribution range in Montenegro (Figure 1). The sampling sites were Orjen (ORJ), Prekornica (PRE), Kučka Korita (KKO), Kučka korita North (KKN), Kuči MT (KMT) and Prokletije (PRO) (Figure 1, Table 1).

The sampling sites differed in altitude, morphology of terrain and soil properties (Table 1, Figure 2). Sites at ORJ and KMT were on altitudes of ca. 1800 m, with trees growing on stone ridges, high inclinations, and lithosols. Sites at PRE, PRO and KKO were on an altitude of ca. 1250 m, on flat terrain and more developed soil that was leptic cambisol. A site at KKN was in close proximity to the KKO, but it was on a slope with a thinner soil layer (molic leptosol). Trees at the KKN were in groups heavily infested by *Tomicus* sp. bark beetles (Scolitinae). Trees at ORJ and KMT were growing at the upper tree line, and thus, these sites were considered as sites with harsh growing conditions, while other sites had moderate growing conditions.

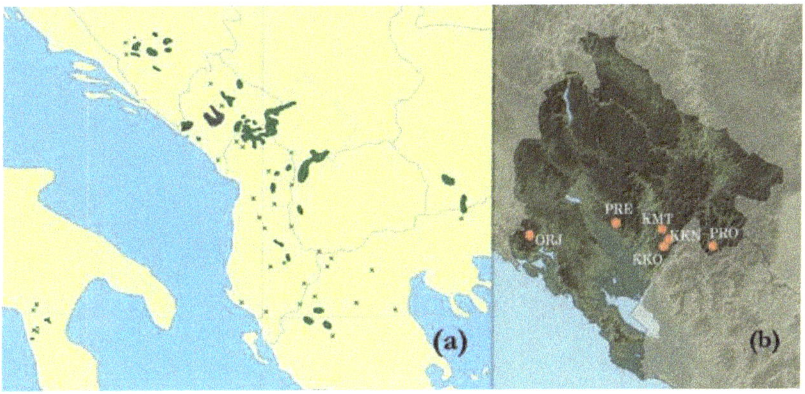

Figure 1. A map showing (**a**) distribution of *Pinus heldreichii* in the Mediterranean region; a grey area was added and represents an additional *P. heldreichii* site; the map is adapted from [18] and (**b**) sampling sites in Montenegro (ORJ-Orjen; PRE-Prekornica; KKO-Kuči Mt, Kučka korita; KKN-Kuči Mt., Kučka korita North, KMT- Kuči mountain, Momonjevo, PRO- Prokletije); a map was adapted from [19].

Table 1. Characteristics of *Pinus heldreichii* sampling sites.

Site	GPS Coordinates	Altitude (m)	Soil Type [a]	Climate [b]	Health [c]
Orjen, Reovačka greda (ORJ)	42°35′ N 18°35′ E	1700	Lithosol	Cfsb	Healthy-looking
Kuči Mt., Momonjevo (KMT)	42°36′ N 19°32′ E	1800	Lithosol	Cfb	Healthy-looking
Prekornica, Studeno (PRE)	42°38′ N 19°12′ E	1200	Leptic cambisol	Cfsb	Healthy-looking
Prokletije, Ropojana (PRO)	42°29′ N 19°48′ E	1250	Leptic cambisol	Cfb	Healthy-looking
Kučka korita (KKO)	42°29′ N 19°30′ E	1300	Leptic cambisol	Cfs″b	Healthy-looking
Kučka korita North (KKN)	42°30′ N 19°32′ E	1450	Molic leptosol	Cfs″b	Attacked by insects

[a] WRB soil clasification system [20,21], [b] Koopen climate classification [22,23], [c] Health status of the trees.

Figure 2. (**a**) *Pinus heldreichii* shoots with needles; trees growing at (**b**) Kuči Mountains, Momonjevo (KMT), and (**c**) Kuči Mountains, Kučka korita (KKO) sites.

Climate at the study sites is characterised as a humid warm temperate climate type (Cf), which is represented by subtypes Cfsb, Cfs″b and Cfb [22,23]. Summers are short, dry and chilly, and winters are cold and windy. At different study sites, the mean annual air temperature is between 1 °C (at PRO) and 6 °C (at ORJ). The mean daily summer maximum is between 7 and 11 °C, and the mean daily winter minimum is between −5 and 0 °C. The absolute minimum is in winter at ca. −30 °C and the maximum is in summer, at ca. 30 °C. The mean annual precipitation is highest at ORJ with ca. 3800 mm, and lowest at PRO with ca. 2000 mm. The rainfall reaches maximum in late autumn and early winter, while the minimum is during the summer months that is often followed by 40–70 day-long periods of droughts. The mean summer precipitation is 300 mm at ORJ and ca. 220 mm on other sites, making ca. 10% of the total annual precipitation [22].

The bedrock at the study sites is solid chalk limestone, which contains a small proportion of insoluble residues. In general, soils are poorly developed, very water porous, skeletal leptosols (rendzina), having only A horizon, characterized by accumulation of humus (A-R profile). According to the classification of texture, the soil is a sandy loam and the structure is powdery. Soil is faintly differentiated down in the soil profile. Soil is rich in humus content (10–25%), poor in calcium-carbonate and has a weak acid reaction (pH ca. 6). On more flat terrains, leptosol develops into the leptic cambisol, which contains initial B horizon. Brown B horizon improves the water retention capacity of soil [20,24].

2.2. *Experimental Design and Sampling*

At the sampling sites, stands of *P. heldreichii* were healthy-looking (Figure 2a). An exception was the KKN site, where trees were damaged by *Tomicus* sp. bark beetles, resulting in discolouration and fungal infection of needles. At all sites, the shoot dieback was occasional, but with higher rates at the KKN site.

Sampling was carried out in May 2015, i.e., before the beginning of the growing season. At each site, experimental design included sampling of five twigs with needles from five different trees. Hence, at each site, five mature *P. heldreichii* trees situated at a distance of ca. 50 m from each other were selected and five twigs of up to 15 cm long and up to 2 m from the ground were sampled from different parts of the crown using secateurs. This sampling approach was used in order (i) to get a joint representative sample per each site (all samples per site were amplified with the same barcode, see below); (ii) to compare fungal communities among different sites, but not within the same site. Collected twigs were placed in plastic bags and transported to the laboratory.

In the laboratory, twigs and needles were assessed for the presence of disease symptoms. The current-year needles were typically green and without disease symptoms. Symptomatic needles were generally two years old or older and symptoms included changed colour that was yellow, orange, pale green or yellowish with red bands. Other symptoms were dying needle tips and a necrotic base of needles. In the case of insect attack (KKN site), defoliation and necrotic needles were common. In order to sample the potentially entire fungal community associated with needles of *P. heldreichii*, both healthy looking needles and needles with disease symptoms were selected. No surface sterilization was carried out. Following morphological examination, up to 5 representative needles that were 2–4 years old, were selected per twig. Selected needles were cut into 0.5–1 cm long segments, containing a random mixture of both asymptomatic and symptomatic parts of needles, placed in 2 mL screw cap tubes, and stored frozen at −20 °C before DNA extraction.

2.3. *DNA Isolation, Amplificationand Sequencing*

DNA extractions were done from 150 samples (6 sites × 5 trees × 5 needle samples), which previously were freeze-dried for 48 h. For isolation of total DNA, needles were homogenised in a Fastprep machine (Precellys, Montigny-le-Bretonneux, France). The extraction was completed using CTAB protocol [25]. After extraction, DNA samples were purified using a JetQuick DNA purification kit (Genomed GmbH, Leinfelden, Germany). The DNA concentration of each sample was determined using a NanoDrop™ One spectrophotometer (Thermo Scientific, Rodchester, NY, USA) and adjusted to 1–10 ng/μL. Amplification by PCR of the ITS2 rDNA region was done using barcoded fungal-specific primer gITS7 [26] and barcoded universal primer ITS4 [27]. All 25 samples from the same site were amplified using primers with the same barcode, resulting in six different barcodes representing each site. Amplification of several samples with the same barcode was done to get a broader representativeness of fungal communities per site. Amplifications were performed using the Applied Biosystems 2720 thermal cycler (Foster City, CA, USA). An initial denaturation step started at 95 °C for 2 min, followed by 27 amplification cycles of denaturation at 95 °C for 30 s, annealing at 55 °C for 30 s, and extension at 72 °C for 60 s. The thermal cycling was ended by a final extension step at 72 °C for 7 min. The PCR products were analyzed using gel electrophoresis on 1% agarose gels stained with Nancy-520 (Sigma-Aldrich, Sweden). PCR products were purified using a sodium acetate protocol [28]. Purified PCR products were quantified using a Qubit fluorometer 4.0.(Thermo Fisher Scientific, Waltham, MA, USA), and an equimolar mix of all PCR products with six barcodes (one per each site) was used for high-throughput sequencing using a Pacific Biosciences platform (Menlo Park, CA, USA) at the SciLifeLab (Uppsala, Sweden).

2.4. Bioinformatics

Principles of bioinformatics followed that specified in [29]. The sequences obtained from the six samples that represented six sampling sites (Table 2), were subjected to quality control and clustering in the SCATA NGS sequencing pipeline [30]. The initial procedure started with quality filtering of the sequences that included the removal of sequences shorter than 200 bps, sequences with low read quality, primer dimers and homopolymers, which were collapsed to 3 bps before clustering. Only sequences containing a barcode and primer were retained. Then, the primer and sample barcodes were removed from the sequence, but information on the sample and sequence association was stored as meta-data. A single-linkage clustering based on 98% similarity was used to cluster sequences into different taxa. For each cluster, the sequence of the most common genotype was used for taxonomic identification. For clusters containing only two sequences, a consensus sequence was produced. The taxa were taxonomically identified using the GenBank database and the Blastn algorithm [31]. The following criteria were used for identification: sequence coverage >80%; 94–97% similarity to genus level and >98% similarity to species level. Sequences deviating from these criteria were identified only to a high taxonomic rank and were given unique names as shown in Table 3 and Table S1. Representative sequences of fungal nonsingletons are available from GenBank under accession numbers MT241905–MT242268.

Table 2. Generated high-quality ITS2 rDNA fungal sequences and detected diversity of fungal taxa at different sampling sites of *Pinus heldreichii*. Within the column *No. of Fungal Taxa*, values followed by the same letter in chi-square test do not differ significantly at $p > 0.05$.

Site	No. of Fungal Sequences	No. of Fungal Taxa	Shannon Diversity Index
Orjen –ORJ	4830	139 a	3.4
Kuči Mt.-KMT	4910	119 ab	2.9
Prekornica-PRE	5828	192 a	3.1
Prokletije -PRO	3679	127 a	3.2
Kučka korita -KKO	8748	167 b	2.9
Kučka korita North –KKN	3834	104 ab	2.7
All	31,829	375	-

Table 3. Relative abundance of the 25 most common fungal taxa detected in needles of *Pinus heldreichii* at six sites in Montenegro. Sites are as in Table 1.

Fungal Taxon	Phylum *	GenBank Accession No.	Compared (bp)	Similarity (%)	Sites						All
					ORJ	KMT	PRE	PRO	KKO	KKN	
Lophodermium pinastri	A	KC608049	361/361	100	4.7	22.1	14.3	14.2	14.9	0.4	12.5
Lophodermium conigenum	A	HM060650	453/453	100	-	0.1	32.4	5.5	14.5	2.8	10.9
Sydowia polyspora	A	KU516591	473/473	100	6.8	16.0	2.1	8.2	0.4	32.2	8.8
Cyclaneusma niveum	A	AF013223	442/442	99	1.3	1.0	4.5	5.7	12.9	0.9	5.5
Unidentified sp. 2814_1	A	MF976656	238/245	97	20.2	-	0.5	-	8.2	-	5.4
Phaeosphaeria pontiformis	A	KT000144	442/442	98	-	-	3.7	11.1	8.9	-	4.4
Unidentified sp. 2814_11	A	KU062806	203/252	81	-	-	1.3	9.0	10.3	-	4.1
Unidentified sp. 2814_10	A	KX220267	257/257	100	3.3	-	1.5	4.1	5.4	5.4	3.4
Phaeomoniella sp. 2814_15	A	GQ153187	257/259	99	7.2	5.9	0.2	1.2	2.9	0.5	3.0
Neocatenulostroma germanicum	A	KR995100	242/242	100	1.3	-	0.4	1.0	3.4	7.9	2.3
Microsphaeropsis olivacea	A	MH871969	249/249	100	0.7	0.1	0.3	0.1	-	16.8	2.2
Allantophomopsiella pseudotsugae	A	MH857222	240/240	100	6.4	7.1	-	-	-	-	2.1
Lachnellula calyciformis	A	MH858771	239/239	100	4.7	8.1	-	-	-	-	2.0
Unidentified sp. 2814_23	A	KF983527	209/244	86	1.4	-	1.5	3.7	0.3	4.3	1.5
Ramoconidiophora euphorbiae	A	MG592740	232/239	97	3.2	6.0	-	-	-	0.6	1.5
Geastrumia sp. 2814_26	A	FJ438389	220/235	94	2.7	2.6	0.4	0.6	0.5	2.3	1.4
Unidentified sp. 2814_16	A	KP892077	240/242	99	4.8	0.1	0.6	1.9	0.3	0.5	1.2
Unidentified sp. 2814_27	A	KT244857	209/248	84	-	-	3.5	5.0	-	-	1.2
Collophora sp. 2814_30	A	NR_137726	229/241	95	3.2	2.2	0.4	-	-	0.3	0.9
Unidentified sp. 2814_29	A	MF976139	209/240	87	1.4	-	0.4	4.1	0.4	-	0.9
Rhytismataceae sp. 2814_22	A	KR266446	234/237	99	0.4	-	1.0	1.4	1.6	-	0.8
Dothideomycetes sp. 2814_13	A	KP991484	253/257	98	1.1	3.5	-	0.1	-	0.4	0.8
Mollisia ligni	A	MF161301	237/241	98	-	4.7	-	-	-	-	0.7
Cenangium acuum	A	MG597445	239/239	100	-	-	3.9	-	-	-	0.7
Athelia acrospora	B	KP814375	296/299	99	-	-	1.8	0.1	1.3	-	0.7
All of 25 taxa					74.9	79.9	74.6	76.6	86.4	75.4	79.0

* A—Ascomycota, B—Basidiomycota.

2.5. Statistical Analyses

Rarefaction analysis was carried out using Analytical Rarefaction v.1.3 [32]. Differences in richness of fungal taxa in different study sites of *P. heldreichii* were compared by nonparametric chi-square testing [33]. As each of the datasets was subjected to multiple comparisons, confidence limits for *p*-values of chi-square tests were reduced the corresponding number of times as required by the Bonferroni correction [34]. The Shannon diversity index, qualitative Sorensen similarity index and nonmetric multidimensional scaling (NMDS) in Canoco 5 [33,35,36] were used to characterize the diversity and composition of fungal communities.

3. Results

The phyllosphere fungi of *P. heldreichii* were examined from four different mountain regions, represented by six sampling sites across the distribution range of *P. heldreichii* in Montenegro (Figure 1). Amplification of fungal ITS2 rDNA from 150 needle samples, PacBio sequencing and quality filtering resulted in 31,829 high quality reads. Sequence assembly and BLASTn analyses showed that the fungal community in the phyllosphere of *P. heldreichii* was composed of 375 fungal taxa (Table 2, Table S1). Nonfungal taxa and singletons were excluded. The detected fungi were 295 (78.7%) Ascomycota, 79 (21.0%) Basidiomycota and 1 (0.2%) Mortierellomycotina. Identification at least to genus level was successful for 254 (67.7%) out of 375 fungal taxa (Table S1), and those represented 78.5% of all fungal sequences.

In different study sites, the number of fungal taxa varied between 104 and 192 (Table 2). The chi-square test showed that the largest difference in richness of fungal taxa was between KKO and the remaining sites (Table 2). Rarefaction showed that fungal taxa detected in all sites did not reach the species saturation (Figure 3).

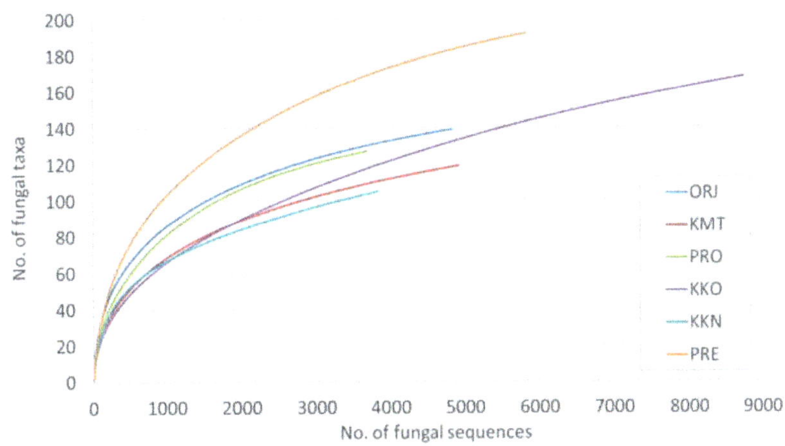

Figure 3. Rarefaction curves showing the relationship between the cumulative number of fungal taxa and the number of ITS2 rDNA sequences from needles of *Pinus heldreichii* from six different sites.

Information on the 25 most common fungal taxa representing 79.0% of all fungal sequences is in Table 3. The most common fungi in the phyllosphere of *P. heldreichii* were *Lophodermium pinastri* (12.5% of all fungal sequences), *Lophodermium conigenum* (10.9%), *Sydowia polyspora* (8.8%), *Cyclaneusma niveum* (5.5%) and Unidentified sp. 2814_1 (5.4%) (Table 3). The most common fungal pathogens of pine needles were *S. polyspora* (8.9%), *C. niveum* (5.5%), *Neocatenulostroma germanicum* (2.3%), *Allantophomopsiella pseudotsugae* (2.1%) and *Cenangium acuum* (0.7%) (Table 3). A pine needle pathogen, *Dothistroma septosporum* (0.25%) was also detected, but at lower relative abundance (Table S1). The other detected pathogenic species that are mainly known as wound pathogens of deciduous trees and/or agricultural crops, were *Phaeomoniella* 2814_15 (3.0%), *Ramoconidiophora euphorbiae* (1.5%), *Collophorina sp.* (0.9%), *Geastrumia sp.* (1.4%) and *Athelia acrospora* (0.7%). The detected fungal endophytes that are known to produce antimicrobial metabolites were *Phaeosphaeria pontiformis* (4.4%), *Microsphaeropsis olivacea* (2.2%), *Lachnellula calyciformis* (2.0%) and *Mollisia ligni* (0.7%). Ubiquitous saprotrophs were rare (Table S1).

The community composition of the phyllosphere fungi varied among different sites (Figure 4). Dothideomycetes dominated fungal communities at PRO and KKN sites, while Leotiomycetes dominated at KMT and PRE sites (Figure 4). At ORJ and KKO sites, the relative abundance of these two classes was similar (Figure 4). Sordariomycetes showed higher relative abundance at ORJ (22.6%), while all other fungal classes in different sampling sites were less abundant (Figure 4). Agaricomycetes (Basidiomycotina) were rare and their relative abundance at different sites varied between 0.03% and 6.3% (Figure 4).

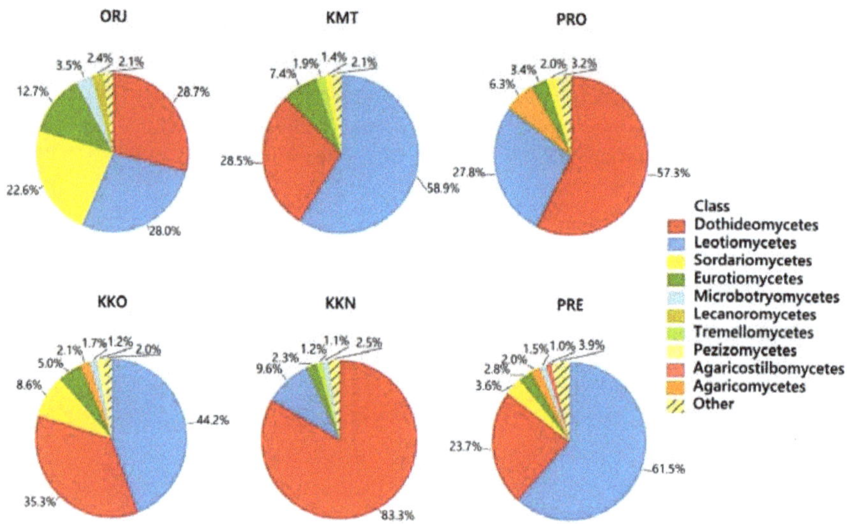

Figure 4. Relative abundance of different fungal classes in needles of *Pinus heldreichii* from six different sites in Montenegro. Classes comprising <1.0% of relative abundance for a given set of sequences were combined and shown as other. Sites are as in Table 1.

The NMDS of fungal communities associated with needles of *P. heldreichii* (Figure 5) showed that KKO, PRO and PRE sites clustered together and were separated from the remaining sites along axis 1. Fungal communities at the ORJ and KMT showed a closer proximity (Figure 5), while fungal communities at the KKN site differed from all remaining sites and were separated along both axis 1 and 2.

At ORJ and KMT sites that are characterized by harsh growing conditions, the fungal community was dominated by *Lophodermium pinastri, Sydowia polyspora, Phaeomoniella* sp., *Allanthopomopsiella pseudotsugae, Lachnellula calyciformis, Ramoconidiophora euphorbiae, Collophorina* sp., *Geastrumia* sp. and *Mollisia ligni* (Table 3). At PRE, PRO and KKO sites that are characterized by moderate growing conditions, fungal community was dominated by *Lophodermium conigenum, L. pinastri, Cyclaneusma niveum, Phaeosphaeria punctiformis,* Unidentified sp. 2814_11, *Cenangium acuum* and *Dothistroma septosporum.* At the KKN, with insect attacks and partly necrotic needles, fungal community was dominated by *S. polyspora, Microsphaeropsis olivacea, Neocatelunostroma germanicum,* Uncultured sp. 2814_10, *Chaetothyriales* 2814_18 and *Geastrumia* sp. (Table S1).

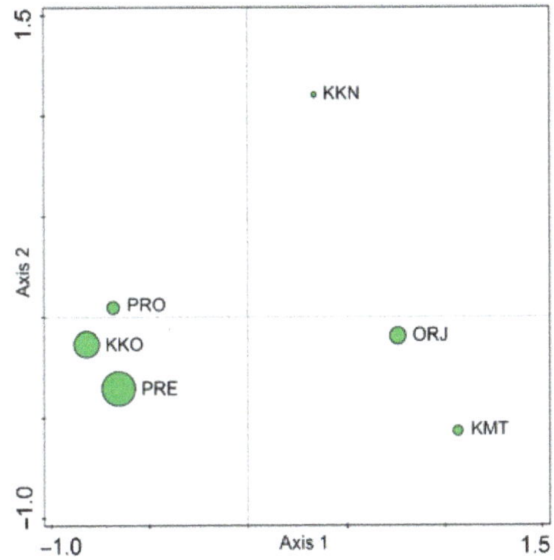

Figure 5. Ordination diagram based on nonmetric multidimensional scaling (NMDS) of fungal communities associated with needles of *Pinus heldreichii* from six different sites in Montenegro, with 56.5% variation on Axis 1 and 28.2% on Axis 2. Each point in the diagram represents a single site and the size of the point reflects richness of the fungal taxa as in Table 2. Sites are as in Table 1.

The Shannon diversity index of fungal communities in different study sites was between 2.7 and 3.4 (Table 2). Among different sites, Sørensen similarity index of fungal communities was low to moderate as it ranged between 0.28 and 0.54 (Table S2).

4. Discussion

A recent development of high-throughput sequencing methods provides powerful tools to explore fungal diversity. Such tools enable identification of complex fungal communities and individual community components directly from environmental samples. Besides, while providing detailed and semi-quantitative information, these tools enable studying the effects of different factors on fungal diversity and community composition [29,37,38]. By using PacBio sequencing, we detected nearly 400 fungal taxa, which were associated with needles of *P. heldreichii* including fungi present in very small abundances (Table S1). Many fungal taxa remained unidentified, which remains a major challenge in fungal taxonomy [39]. However, while using high-throughput sequencing, we need be aware of potential limitations that can include methodological biases, limitations of markers and bioinformatics challenges [40,41].

The observed diversity of the phyllosphere fungi can be considered being high and comparable with similar studies on different *Pinus* species. For example, there were 446 and 260 fungal taxa recorded during studies of needle-associated fungi of *Pinus sylvestris* in Sweden and Poland, respectively [39,42]. Furthermore, by using fungal culturing, there were 118 fungal taxa associated with needles of *Pinus taeda* in USA [43], 35 taxa associated with needles of *Pinus halepensis* [44] and *P. sylvestris* in Spain [45]. According to [45], the species diversity of fungal endophytes varied from 21 in *Pinus monticola* to 49 in *Pinus nigra*. At different study sites, the growing conditions of *P. heldreichii* had in general a minor effect of the absolute richness of fungal taxa, which was similar among different sites (Table 2). However, with respect to the number of sequences obtained from samples of each site, lower richness of fungal taxa was at the KKO (Table 2), what was likely due to the dominance of several fungal taxa such as, e.g., *Lophodermium* species, *Cylaneusma niveum*, Unidentified sp. 2814_11, *Phaeosphaeria pontiformis* and Unidentified sp. 2814_1. Several fungal taxa of the present study were recorded for the first time in the Balkan region such as, e.g., pathogens *Neocatelunostroma germanicum* and *Allantophomopsiella pseudotsugae*. The study also revealed a number of fungal pathogens that were previously known from agricultural crops as, e.g., *Phaeomoniella*, *Ramoconidiophora*, *Geastrumia* and *Athelia*. These could have been overlooked in other studies due to latent occurrence [46]. As *P. heldreichii* belongs to a group of forest trees that grows under harsh enviromental conditions, the possibility should not be excluded that both the host and the specific environment influence the composition of associated fungal communities. Studies on plant–endophyte associations in high stress habitats have revealed that at least some fungal endophytes can contribute to stress tolerance of host plants [12]. Indeed, the results of this study revealed site-specific differences in fungal communities associated with needles of *P. heldreichii*. For example, fungal communities were more similar among sites situated at lower altitudes, i.e., with moderate growing conditions (PRE, PRO, KKO) as compared to those sites at higher altitudes, i.e., with harsh growing conditions (ORJ, KMT) (Table 1, Figure 5). Further, the fungal community at the KKN site was largely different from the remaining sites, what was likely due to insect damage to trees, resulting in mainly symptomatic and partly necrotic needles. These observations demonstrate that the composition of fungal communities and their succession in needles of *P. heldreichii* can be determined by different abiotic and biotic factors. Different environmental factors have earlier been shown to play an important role in shaping fungal communities both in the phyllosphere and in the rhizosphere of forest trees [5,42,47,48].

Fungal pathogens represent an important biotic factor that may negatively affect health and growth of forest trees [11]. In the present study, a fungal pathogen *Sydowia polyspora* was among the most dominant fungi (Table 3). *Sydowia polyspora* has a wide geographical range [48] and is common in Europe. The pathogenicity of *S. polyspora* to young conifers (genera *Thuja*, *Abies*, *Tsuga*, *Larix*, *Picea* and *Pinus*) was previously reported [49]. In the Balkan region, it has been reported in forest plantations of *P. nigra* and *P. sylvestris* in Serbia, occurring on needles damaged by drought or frost [50]. It was also detected on needles of *P. halepensis* in Italy and Spain [44,51], *P. sylvestris* in Poland and Lithuania [29,39] and on *P. ponderosa* in North America [52]. *Sydowia polyspora* has recently been reported as a pathogen that dominates fungal communities vectored by bark beetles associated with *P. radiata*, *P. nigra* subsp. *salzmannii* and *P. sylvestris* in Spain [53] and *Pinus yunnanensis* in China [54]. *Sydowia* symptoms include needle necrosis and shoot dieback. The fungus is favoured by a warm climate, especially if the host is stressed by summer drought or insect or mite attack [53]. Indeed, in the present study the highest relative abundance of *S. polyspora* was at the KKN (Table 2), the site that was subjected to insect attack. Besides, it appears that *S. polyspora* had a negative effect on the relative abundance of *L. pinastri* (Table 2), which is in agreement with Behnke-Borowczyk et al. [39], who have observed the similar pattern on *P. sylvestris*. *Cyclaneusma niveum* was also among the dominant fungi (Table 3). It is one of the two fungal pathogens causing Cyclaneusma needle cast, which is an important needle disease reported from many pine species including *P. nigra* and *P. sylvestris* in the Balkan region and in Crimea [50,55,56], and reported from *P. halepensis* in Spain [44]. It was suggested that *C. niveum* is more frequently associated with *P. nigra* under warmer climate conitions, while *Cyclaneusma minus* is favoured by wet, humid, above-freezing conditions, and thus, more commonly infects *P. sylvestris* grown in central and northern Europe [39,50,57]. Our study is in agreement, as on *P. heldreichii*, *C. niveum* dominated the fungal community, while *C. minus* was recorded at a low frequency (Table S1). Interestingly, *C. niveum* was more common on sites with prevailing moderate than harsh enviromental conditions. *Neocatenulostroma germanicum* is a recently identified fungal pathogen causing needle blight on *P. mugo*, *P. sylvestris* and *P. nigra* subsp. *pallsiana* in Lithuania, Ukraine and Poland, where it was found to be commonly associated with *Dothistroma*, *Lecanosticta acicola* and Cyclaneusma needle cast infections [39,58]. In this study, it was present in the needles of *P. heldreichii* from five sites, with considerably higher relative abundance on symptomatic needles at the KKN site and the neighbouring KKO site as compared to the remaining sites. *Allantophomopsiella pseudotsugae* (syn. *Phomopsis pseudotsugae*) was a pathogen commonly detected on needles of *P. heldreichii* grown on sites under harsh growth conditions (ORJ and KMT) (Table 3). While being a pathogen of conifers, that is, mainly infecting pines, it develops on young shoots [59] and has been reported in several European countries [60]. The present finding of the fungus represents a new record for this part of Europe, indicating that *A. pseudotsugae* could be among important pathogens of *P. heldreichii* growing under harsh conditions. *Cenangium accum* was recorded on *P. heldreichii* needles at the PRE site (Table 3). Previously it was demonstrated as a pathogen of weakened *P. nigra* and *P. sylvestris* in the Balkan region [50,55]. It has also been shown to be associated with Cyclaneusma needle cast in *P. sylvestris* in western Poland [39]. *Cenangium accum* develops predominantly on needles damaged by frosts or drought and its development is favoured by a high humidity. Similarly, *Dothistroma septosporum*, was recorded at a low relative abundance and only on the PRE (Table S1). Investigation of *Dothistroma septosporum* accomplished using PCR and species-specific primers, has demonstrated the presence of this potentially invasive pathogen across the *P. heldreichii* distribution range in Montenegro [17]. It was suggested that enviromental conditions present at *P. heldreichii* sites suppress the development of Dothistroma needle blight and that *P. heldreichii* is only slightly susceptible to *D. septosporum*. However, the infection level by *D. septosporum* may vary in different years, especially after rainy periods [10,17].

Among the other fungi detected, there were fungal endophytes that are known to be commonly associated with different tissues of forest trees [61,62]. The relationship between endophytic fungi and plants is not clearly understood and may change depending on the health status of the plant. However, for at least part of their life, they colonise plant tissues asymptomatically [12,63,64]. In the present study, *L. pinastri* and *L. conigenum* were the most commonly detected fungi on *P. heldreichii* needles. They are globaly distributed and commonly associated with pines [42,65]. It was shown recently that *L. pinastri* colonises healthy needles latently as an endophyte, initiates active growth at the begining of needle senescence and sporulates after needle fall. It is a dominant coloniser of dying needles and a saprotroph contributing to a needle decomposition [39]. In the Balkan region, *L. pinastri* was frequently reported from *P. nigra* and *P. sylvestris* grown in forests, forest nurseries and plantations [50,55]. *Lophodermium conigenum* is known as a coloniser of damaged needles and can also form fruitbodies on broken branches [65]. In comparison to *L. pinastri*, it is less frequently reported on both *P. nigra* and *P. sylvestris* in Serbia [50] and on *P. sylvestris* in Spain [45] and the UK [66]. In the present study, *L. pinastri* was present on all investigated sites, while *L. conigenum* showed a higher relative abundance on sites characterised by moderate growth conditions (PRO, KKO and PRE). The relative abundance of *L. pinastri* was at least 10 times lower on the KKN site than on any other site, what was likely due to its exclusion by pathogenic fungi which dominated on this site.

Some fungal endophytes of conifers can produce bioactive secondary metabolites, supporting ecological adaptations of host plants owing strong antimicrobial activities [63,67]. *Microsphaeropsis olivacea* (syn. *Coniothyrium olivaceum*) is a fungal endophyte, which was shown to produce bioactive compounds that are considered as promising antibacterial and antifungal agents [68]. *Microsphaeropsis olivacea* was among the most abundant species on the KKN, where *Sydowia polyspora* and *Neocatenulostroma germanicum* were also abundant. *Mollisia ligni* has been shown to produce mollisin, a compound that is known for a strong fungicidal activity against *Sydowia polyspora* [69]. A high frequency of *Mollisia ligni* may result in decreased frequency of *S. polyspora* on *P. sylvestris* [39]. On *P. heldreichii*, *N. germanicum* was absent in the cases where *M. ligni* was present (KMT).

The study has also detected a number of fungi previously known as pathogens of deciduous trees and/or of agricultural crops including *Phaeomoniella* 2814_15, *Ramoconidiophora euphorbiae* and *Geastrumia sp.*, which were for the first time recorded on conifers, possibly as latent endophytes. Fungi from genus *Phaeomoniella* are important pathogens of grapevines, causing grapevine trunk disease [70], *Ramoconidiophora* and *Collophorina* species have been reported from necrotic and symptomless wood and leaves of *Prunus*, *Castanea*, *Vitis*, and from roots of *Calluna* [71], while *Geastrumia sp.* is known as one of the species involved in disease complex of sooty blotch and flyspeck of apple (*Malus domestica*) [72]. *Phaeomoniella sp.*, *Ramoconidiophora*, *Collophorina* and *Geastrumia* were more abundant on sites with harsh growth conditions (ORJ and KMT). Among other pathogens, *Lachnellula calyciformis* was abundant on sites under harsh growing conditions, while *Phaeosphaeria pontiformis* was on sites characterised by moderate growing conditions. Corticoid fungus from genus *Athelia*, which typically occurs as saprotroph [73,74] was present on sites with moderate growth conditions.

Fungi may modulate stress tolerance, enhance growth and increase reproduction [7,47,75]. Habitats subjected to a high abiotic stress can be inhabited by a species-specific fungal community [76,77]. Our results demonstrated that needles of *P. heldreichii* constitute a habitat for a species-rich community of fungi, the composition of which was found to be largely driven by environmental conditions and/or health status of host trees. Indeed, similar patterns were also observed for fungal communities in roots of *P. heldreichii* [5], indicating that without the habitat-adapted fungal symbionts, plants are hardly capable of surviving in high stress habitats.

Fungal communities evolve together with host plants [12], suggesting that relic and endemic pine species such as *P. heldreichii* [5,54], can be associated with a specific fungal community. For example among the 50 mushroom species, which were recorded in *P. heldreichii* forests to date [78–81], many were rarely observed and exclusively recorded in *P. heldreichii* forests. Hence, *Chalciporus ammarelus*, *Geastrum minimum*, *Hygroporus gliocyclus*, *Hygroporus hypothejus*, *Rhizopogon roseolus* and *Morchella esculenta* s.l. are listed on the preliminary red list [79], and thus, are protected [82,83]. Furthermore, recently described species *Erioscyphella curvispora* [84] was discovered in *P. heldreichii* needle litter and a newly described genus *Perzia*, accommodated by a type species *Perzia triseptata*, gen. nov. was discovered on the xeric bark of *P. heldreichii* [85]. Other findings include newly described species *Velutarina bertinscensis* [86], *Peziza montrivicola* [87], *Cenangiopsis ragvanii*, *C. junipericola* [88] and rare species *Trichophaea flavobrunea* [89] that were determined on other tree species growing on *P. heldreichii* sites. Those findings represents a special value to science, and together with overall recorded fungal diversity demonstrate a high value of *P. heldreichii* habitats as being unique and biodiversity hotspots in high-altitude mountain areas.

In summary, the results demonstrated that needles of *P. heldreichii* were associated with a species-rich community of fungi, the composition of which was found to be largely dependent on environmental conditions and/or health status of host trees. Needle pathogens appeared to be an important component of fungal communities associated with needles of *P. heldreichii*, but caused limited damage likely due to interaction with other fungal species and/or due to plant defence mechanisms. In order not to become a threat to the health and growth of forest trees, the present and potentially new invasive pathogens should be monitored regularly. Further research is also needed to understand patterns of different fungal species coexistence.

Supplementary Materials: The following are available online at http://www.mdpi.com/1424-2818/12/5/172/s1, Table S1: Relative abundance of fungal taxa detected in needles of *Pinus heldreichii* at six different sites in Montenegro, Table S2: Sørensen similarity index of the phyllosphere fungal communities among the six *Pinus heldreichii* sampling sites.

Author Contributions: Conceptualisation, J.L. and A.M.; methodology, J.L. and A.M.; formal analysis, J.L. and A.M.; resources, J.L.; data curation, J.L. and A.M.; writing—original draft preparation, J.L.; writing—review and editing, J.L. and A.M.; funding acquisition, J.L. All authors have read and agreed to the published version of the manuscript.

Funding: This research was funded by the Ministry of Science of Montenegro through the grant INVO HERIC No: 01-646. A.M. was supported by the Swedish Research Council Formas (grant no. 2019-00597).

Conflicts of Interest: The authors declare no conflict of interest.

References

1. Janković, M.M. Betrachtungen uber gegenseitigen Bezichungen der Molika (*Pinus peuce*) und Panzerkiefer (*Pinus heldreichii*) sowie auch Uber ihre okologishen Eigenschaften, besonders in Bezug auf ihre geologishe Grundlage. *Bull. L'Inst. Jard. Bot. L'univ. Beogr.* **1960**, *1*, 141–181.
2. Jovanović, B. *Dendrology*, 6th ed.; Faculty of Forestry, University of Belgrade: Belgrade, Serbia, 2007; p. 536.
3. Blečić, V.; Lakušić, R. Forests of Pinius heldreichii Christ at Štitovo and Bjelasica in Montenegro. *Bull. Repub. Inst. Prot. Nat. Mus. Nat. Hist. Titogr.* **1969**, *4*, 5–10.
4. Stevanović, V.; Jovanović, S.; Lakušić, D. Diversity of vegetation of Yugoslavia. In *Biodiversity of Yugoslavia with List of Species of Special Importance*; Radović, I., Angelus, J., Eds.; Yugoslavia: Ecolibri, Beograd, 1995; pp. 219–241.

5. Lazarević, J.; Menkis, A. Fungi inhabiting fine roots of *Pinus heldreichii* in the Montenegrin montane forests. *Symbiosis* **2018**, *74*, 189–197. [CrossRef]
6. Vendramin, G.G.; Fineschi, S.; Fady, B. *EUFORGEN Technical Guidelines for Genetic Conservation and Use for Bosnian Pine (Pinus heldreichii)*; Bioversity International: Rome, Italy, 2008; ISBN 978-92-9043-789-5.
7. Baldrian, P. Forest microbiome: Diversity, complexity and dynamics. *Fems Mycrobiol. Rev.* **2017**, *41*, 109–130. [CrossRef]
8. Stewart, J.E.; Kim, M.S.; Klopfenstein, N.B. Molecular genetic approaches toward understanding forest-associated fungi and their interactive roles within forest ecosystems. *Curr. For. Rep.* **2018**, *4*, 72–84. [CrossRef]
9. Zanne, A.E.; Abarenkov, K.; Afkhami, M.E.; Aguilar-Trigueros, C.A.; Bates, S.; Bhatnagar, J.M.; Busby, P.E.; Christian, N.; Cornwell, W.K.; Crowther, T.W.; et al. Fungal functional ecology: Bringing a trait-based approach to plant-associated fungi. *Biol. Rev.* **2020**, *95*, 409–433. [CrossRef]
10. Drenkhan, R.; Tomešová-Haataja, V.; Fraser, S.; Bradshaw, R.; Vahalik, P.; Mullett, M.; Martín-García, J.; Bulman, L.; Wingfield, M.; Kirisits, T.; et al. Global geographic distribution and host range of *Dothistroma* species: A comprehensive review. *For. Pathol.* **2016**, *46*, 408–442. [CrossRef]
11. Elvira-Recuenco, M.; Cacciola, S.O.; Sanz-Ros, A.V.; Garbelotto, M.; Aguayo, J.; Solla, A.; Mullett, M.; Drenkhan, T.; Oskay, F.; Aday Kaya, A.G.; et al. Potential Interactions between Invasive *Fusarium circinatum* and Other Pine Pathogens in Europe. *Forests* **2020**, *11*, 7. [CrossRef]
12. Rodriguez, R.J.; Redman, R.S. More than 400 million years of evolution and some plants still can't make it on their own: Plant stress tolerance via fungal symbiosis. *J. Exp. Bot.* **2008**, *59*, 1109–1114. [CrossRef]
13. Terhonen, E.; Blumenstein, K.; Kovalchuk, A.; Asiegbu, F.O. Forest tree microbiomes and associated fungal endophytes: Functional roles and impact on forest health. *Forests* **2019**, *10*, 42. [CrossRef]
14. Lazarević, J. Contribution to the study of physiological characteristics of the pathogenic fungus *Herpotrichia juniperi*. *Agric. For.* **2003**, *49*, 95–109.
15. Lazarević, J. *Pinus heldreichii* as the host of brown felt blight (*Herpotrichia juniperi*) in Montenegro. *Mycol. Monten.* **2004**, *7*, 77–89.
16. Lazarević, J. Contribution to the study of te morphological characteristics of the pathogenic fungus *Herpotrichia juniperi*. *Mycol. Monten.* **2005**, *8*, 137–147.
17. Lazarević, J.; Davydenko, K.; Millberg, H. Dothistroma Needle Blight on High Altitude Pine Forests in Montenegro. *Balt. For.* **2017**, *23*, 294–302.
18. Caudullo, G.; Welk, E.; San-Miguel-Ayanz, J. Chorological maps for the main European woody species. *Data Brief* **2017**, *12*, 662–666. [CrossRef]
19. Dinaric_Alps_map-fr.svg: Sémhur, Relief Map of Montenegro svg. Available online: https://commons.wikimedia.org/wiki/File:Relief_Map_of_Montenegro.svg (accessed on 10 January 2020).
20. Fuštić, B.; Đuretić, G. *Soils of Montenegro*; Biotechnical Institute, University of Montenegro: Podgorica, Montenegro, 2000; pp. 156–172.
21. IUSS Working Group WRB. Descriptions, distribution, use and management of Reference Soil Groups. In *World Reference Base for Soil Resources. International Soil Classification System for Naming Soils and Creating Legends for Soil Maps, Udate 2015*; World Soil Resource Reports No. 106; Food and Agriculture Organization of the United Nations: Rome, Italy, 2015; pp. 144–180.
22. Burić, M.; Micev, B.; Mitrović, L. *Atlas of Climate of Montenegro*; Montenegrin Academy of Science and Arts: Podgorica, Montenegro, 2012; pp. 18–62, 86–102, 126–129.
23. Burić, D.; Ducić, V.; Mihajlović, J. The climate of Montenegro: Modificators and types-part two. *Bull. Serb. Geogr. Soc.* **2014**, *94*, 73–90. [CrossRef]
24. Janković, M.M.; Stefanović, K. Ecological relations between relict (sub)endemic pine species *Pinus heldreichii* and the character of geological substrate and soil in Jugoslavia. *Ekologija* **1971**, *6*, 49–61.
25. Menkis, A.; Ihrmark, K.; Stenlid, J.; Vasaitis, R. Root-associated fungi of Rosa rugosa grown on the frontal dunes of the Baltic Sea coast in Lithuania. *Microb. Ecol.* **2014**, *67*, 769–774. [CrossRef]
26. Ihrmark, K.; Bodeker, I.T.M.; Cruz-Martinez, K.; Friberg, H.; Kubartova, A.; Schenck, J.; Strid, Y.; Stenlid, J.; Brandstrom-Durling, M.; Clemmensen, K.E.; et al. New primers to amplify the fungal ITS2 region—Evaluation by 454-sequencing of artificial and natural communities. *FEMS Microbiol. Ecol.* **2012**, *82*, 666–677. [CrossRef]

27. White, T.J.; Bruns, T.; Lee, S.; Taylor, J. Amplification and direct sequencing of fungal ribosomal RNA genes for phylogenetics. In *PCR Protocols: A Guide to Methods and Applications*; Innis, M.A., Gelfand, D.H., Sninsky, J.J., White, T.J., Eds.; Academic Press, Inc.: San Diego, CA, USA, 1990; pp. 315–322.
28. Menkis, A.; Marčiulynas, A.; Gedminas, A.; Lynikienė, J.; Povilaitienė, A. High-throughput sequencing reveals drastic changes in fungal communities in the phyllosphere of Norway spruce (*Picea abies*) following invasion of the spruce bud scale (*Physokermes piceae*). *Microb. Ecol.* **2015**, *70*, 904–911. [CrossRef]
29. Lynikienė, J.; Marčiulynienė, D.; Marčiulynas, A.; Gedminas, A.; Vaičiukynė, M.; Menkis, A. Managed and unmanaged *Pinus sylvestris* forest stands harbour similar diversity and composition of the phyllosphere and soil fungi. *Microorganisms* **2020**, *8*, 259. [CrossRef]
30. Brandström Durling, M.; Clemmensen, K.; Stenlid, J.; Lindahl, B. SCATA—An Efficient Bioinformatic Pipeline for Species Identification and Quantification after High-Throughput Sequencing of Tagged Amplicons (Submitted). 2011. Available online: https://www.scata.mykopat.slu.se (accessed on 12 November 2019).
31. Altschul, S.F.; Madden, T.L.; Schäffer, A.A.; Zhang, J.; Zhang, Z.; Miller, W.; Lipman, D.J. Gapped BLAST and PSI-BLAST: A new generation of protein database search programs. *Nucleic Acids Res.* **1997**, *25*, 3389–3402. [PubMed]
32. Analytic Rarefaction, UGA Stratigraphy Lab. Available online: http://www.uga.edu/strata/software/index.html (accessed on 10 January 2020).
33. Magurran, A.E. *Ecological Diversity and Its Measurement*; Princeton University Press: Princeton, NJ, USA, 1988; p. 192.
34. Sokal, R.R.; Rohlf, F.J. *Biometry: The Principles and Practice of Statistics in Biological Research*, 4th ed.; H. Freeman & Company: New York, NY, USA, 2005; p. 896.
35. Shannon, C.E. A mathematical theory of communication. *Bell Syst. Tech. J.* **1948**, *27*, 379–423.
36. Ter Braak, C.J.F.; Smilauer, P. *Canoco Reference Manual and User's Guide to Canoco for Windows: Software for Canonical Community Ordination, Version 4*; Microcomputer Power: Ithaca, NY, USA, 1998; p. 351.
37. Bringel, F.; Couee, I. Pivotal roles of phyllosphere microorganisms at the interface between plant functioning and atmospheric trace gas dynamics. *Front. Microbiol.* **2015**, *6*, 486. [CrossRef] [PubMed]
38. Porazinska, D.L.; Sung, W.; Giblin-Davis, R.M.; Thomas, W.K. Reproducibility of read numbers in high-throughput sequencing analysis of nematode community composition and structure. *Mol. Ecol. Resour.* **2010**, *10*, 666–676. [CrossRef] [PubMed]
39. Behnke-Borowczyk, J.; Kwaśna, H.; Kulawinek, B. Fungi associated with *Cyclaneusma* needle cast in Scots pine in the west of Poland. *For. Pathol.* **2019**, *49*, e12487. [CrossRef]
40. Lindahl, B.D.; Nilsson, R.H.; Tedersoo, L.; Abarenkov, K.; Carlsen, T.; Kjøller, R.; Kõljalg, U.; Pennanen, T.; Rosendahl, S.; Stenlid, J.; et al. Fungal community analysis by high-throughput sequencing of amplified markers—A user's guide. *New Phytol.* **2013**, *199*, 288–299.
41. Tedersoo, L.; Tooming-Klunderud, A.; Anslan, S. PacBio metabarcoding of Fungi and other eukaryotes: Errors, biases and perspectives. *New Phytol.* **2018**, *217*, 1370–1385.
42. Millberg, H.; Boberg, J.; Stenid, J. Changes in fungal community of Scots pine (*Pinus sylvestris*) needles along a latitudinal gradient in Sweden. *Fungal Ecol.* **2015**, *17*, 126–139. [CrossRef]
43. Oono, R.; Lefèvre, E.; Simha, A.; Lutzoni, F. A comparison of the community diversity of foliar fungal endophytes between seedling and adult loblolly pines (*Pinus taeda*). *Fungal Biol.* **2015**, *119*, 917–928. [CrossRef]
44. Botella, L.; Santamaria, O.; Diez, J.J. Fungi associated with decline of *Pinus halepensis* in Spain. *Fungal Divers.* **2010**, *40*, 1–11. [CrossRef]
45. Sanz-Ros, A.V.; Muller, M.M.; San Martin, R.; Diez, J.J. Fungal endophytic communities on twigs of fast and slow growing Scots pine (*Pinus sylvestris* L.) in northern Spain. *Fungal Biol.* **2015**, *119*, 870–883. [CrossRef]
46. Parfitt, D.; Hunt, J.; Dockrell, D.; Rogers, H.J.; Boddy, L. Do all trees carry the seeds of their own destruction? PCR reveals numerous wood decay fungi latently present in sapwood of a wide range of angiosperm trees. *Fungal Ecol.* **2010**, *3*, 338–346. [CrossRef]
47. Bowman, E.A.; Arnold, A.E. Distribution of ectomicorrhizal and foliar endophytic fungal communities associated with *Pinus ponderosa* along a spatially constrained elevation gradient. *Am. J. Bot.* **2018**, *105*, 687–699. [CrossRef]
48. Botella, L.; Diez, J.J. Phylogenetic diversity of fungal endophytes in Spanish stands of *Pinus halepensis*. *Fungal Divers.* **2011**, *47*, 9–18.

49. Talgo, V.; Chastanger, G.; Thomsen, I.M.; Cech, T.; Riley, K.; Lang, K.; Klemsdal, S.S. *Sydowia polyspora* associated with current season needle necrosis (CSNN) on true fir (*Abies* spp.). *Fungal Biol.* **2010**, *114*, 545. [CrossRef] [PubMed]
50. Karadžić, D.; Milijašević, T. The most important parasitic and saprophytic fungi in Austrian pine and Scots pine plantations in Serbia. *Bull. Fac. For.* **2008**, *97*, 147–170.
51. Tinivella, F.; Dani, E.; Minuto, G.; Minuto, A. First Report of *Sydowia polyspora* on Aleppo Pine (*Pinus halepensis*) in Italy. *Plant Dis.* **2014**, *98*, 2. [CrossRef]
52. Ridout, M.; Newcombe, G. *Sydowia polyspora* is both a Foliar Endophyte and a Preemergent Seed Pathogen in *Pinus Ponderosa*. *Plant Dis.* **2018**, *118*, 3. [CrossRef]
53. Muñoz-Adalia, E.J.; Sanz-Ros, A.V.; Flores-Pacheco, J.A.; Hantula, J.; Diez, J.J.; Vainio, E.J.; Fernández, M. *Sydowia polyspora* Dominates Fungal Communities Carried by Two *Tomicus* Species in Pine Plantations Threatened by *Fusarium circinatum*. *Forests* **2017**, *8*, 127. [CrossRef]
54. Pan, Y.; Ye, H.; Lu, J.; Chen, P.; Zhou, X.D.; Qiao, M.; Yu, Z.-F. Isolation and identification of *Sydowia polyspora* and its pathogenicity on *Pinus yunnanensis* in Southwestern China. *J. Phytopatol.* **2018**, *166*, 386–395. [CrossRef]
55. Dobreva, M.; Georgieva, M.; Dermedzhiev, P.; Nachev, R.; Velinov, V.; Terziev, P.; Georgiev, G. Fungal pathogens associated with *Pinus species* in the region of forest protection station Plovdiv in the period 2013–2016. *For. Sci.* **2016**, 1–2.
56. Minter, D.W.; Dudka, I.O. *Fungi of Ukraine. A Preliminary Checklist*; International Mycological Institute: Egham, Surrey, UK; M.G. Kholodny Institute of Botany: Kiev, Ukraine, 1996; pp. 1–361.
57. Drenkhan, R.; Hanso, M. Recent invasion of foliage fungi of pines (*Pinus* spp.) to the Northern Baltics. *For. Stud.* **2009**, *51*, 49–64. [CrossRef]
58. Markovskaja, S.; Kačergius, A.; Davydenko, K.; Fraser, S. First record of *Neocatenulostroma germanicum* on pines in Lithuania and Ukraine and its co- occurrence with *Dothistroma* spp. and other pathogens. *Path* **2016**, *46*, 522–533. [CrossRef]
59. Wajihi, A.H.; Lee, S.; Das, K.; Eom, A.; Jung, H. First Report of *Allantophomopsiella pseudotsugae* Isolated from Soil in Korea. *Korean J. Mycol.* **2019**, *47*, 29–34. [CrossRef]
60. Crous, P.W.; Quaedvlieg, W.; Hansen, K.; Hawksworth, D.L.; Groenewald, J.Z. *Phacidium* and *Ceuthospora* (Phacidiaceae) are congeneric: Taxonomic and nomenclatural implications. *Ima Fungus* **2014**, *5*, 173–193. [CrossRef]
61. Sieber, T.N. Endophytic fungi in forest trees: Are they mutualists? *Fungal Biol. Rev.* **2007**, *21*, 75–89. [CrossRef]
62. Kogel, K.-H.; Franken, P.; Huckelhoven, R. Endophyte or parasite-what decides. *Curr. Opin. Plant Biol.* **2006**, *9*, 358–369. [CrossRef]
63. Soltani, J.; Hosseyni Moghaddam, M.S. Fungal Endophyte Diversity and Bioactivity in the Mediterranean Cypress *Cupressus Sempervirens*. *Curr. Microbiol.* **2015**, *70*, 580–586.
64. Jeewon, R.; Yeung, Q.S.; Wannasinghe, D.N.; Rampadarath, S.; Puchooa, D.; Wang, H.K.; Hyde, K.D. Hidden mycota of pine needles: Molecular signatures from PCRDGGE and Ribosomal DNA phylogenetic characterization of novel phylotypes. *Sci. Rep.* **2018**, *8*, 18053. [CrossRef]
65. Minter, D.W.; Staley, J.M.; Millar, C.S. Four species of *Lophodermium* on *Pinus sylvestris*. *Trans. Br. Mycol. Soc.* **1978**, *71*, 295–301.
66. Reignoux, S.A.; Green, S.; Ennos, R.A. Molecular identification and relative abundance of cryptic *Lophodermium* species in natural populations of Scots pine, *Pinus sylvestris* L. *Fungal Biol.* **2014**, *118*, 835–845. [CrossRef] [PubMed]
67. Stierle, A.A.; Stierle, D.B. Bioactive secondary metabolites produced by fungal Endophytes of conifers. *Nat. Prod. Commun.* **2015**, *10*, 1671–1682. [PubMed]
68. Wang, L.W.; Xu, B.G.; Wang, J.Y.; Su, Z.Z.; Lin, F.C.; Zhang, C.L.; Kubicek, C.P. Bioactive metabolites from Phoma species, an endophytic fungus from the Chinese medicinal plant Arisaema erubescens. *Appl. Microbiol. Biotechnol.* **2012**, *93*, 1231–1239. [CrossRef] [PubMed]
69. Johne, S.; Watzke, R.; Meusel, W.; Mollmann, V.; Harti, A.; Dahse, H.M.; Matthes, B.; Seifert, K. Biotechnological production and bioactivities of mollisin and two new, structuraly related fungal naphthequinone metabolites. *Chem. Biodivers.* **2005**, *2*, 1109–1115. [PubMed]

70. Bertscha, C.; Ramırez-Sueroa, M.; Magnin-Robertb, M.; Larignonc, P.; Chonga, J.; Abou-Mansourd, E.; Spagnolob, A.; Clementb, C.; Fontaineb, F. Grapevine trunk diseases: Complex and still poorly understood. *Plant Pathol.* **2013**, *62*, 243–265. [CrossRef]
71. Bien, S.; Kraus, C.; Damm, U. Novel *Collophorina*-like genera and species from *Prunus* trees and vineyards in Germany. *Persoonia* **2020**, *45*, 46–67. [CrossRef]
72. Piperkova, N.; Yonkova, I. Symptoms, Etiology and Control of Sooty Blotch and Flyspeck in Bulgaria. *Turk. J. Agric. Nat. Sci.* **2014**, *1*, 817–822.
73. Epidemiology of the Fungus *Athelia arachnoidea* in Epiphytic Communities of Broadleaved Forests under Strong Anthropogenic Impact. Available online: http://www.elibrary.lt/resursai/LMA/Ekologija/0504_07_Eko.pdf (accessed on 5 March 2020).
74. Adams, G.C.; Kropp, B.R. *Athelia arachnoidea*, the sexual state of *Rhizoctonia carotae*, a pathogen of carrot in cold storage. *Mycologia* **1996**, *88*, 459–472. [CrossRef]
75. Menkis, A.; Vasiliauskas, R.; Taylor, A.F.S.; Stenlid, J.; Finlay, R. Fungal communities in mycorrhizal roots of conifer seedlings in forest nurseries under different cultivation systems, assessed by morphotyping, direct sequencing and mycelial isolation. *Mycorrhiza* **2005**, *16*, 33–41.
76. Gehring, C.A.; Theimer, T.C.; Whitham, G.; Keim, P. Ectomycorrhizal fungal community structure of pinyon pines growing in two environmental extremes. *Ecology* **1998**, *79*, 1562–1572.
77. Rosa, L.H.; Vaz, A.B.M.; Caligiorne, R.B.; Campolina, S.; Rosa, C.A. Endophytic fungi associaed with Antarctic grass *Deschampsia antarctica* Devs (Poaceae). *Polar Biol.* **2009**, *32*, 161–167. [CrossRef]
78. Perić, B.; Perić, O. Diversity of macromycetes in Montenegro. *MASA* **1997**, *11*, 45–142.
79. Perić, B.; Perić, O. Preliminary red list of Macromycetes of Montenegro 2. *Mycol. Monten.* **2004**, *7*, 7–33.
80. Lazarević, J.; Perić, O.; Perić, B. Ectomycorrhizal fungi in Montenegro—Diversity and distribution. *Mycol. Monten.* **2011**, *14*, 85–115.
81. Lazarević, J. Forests and biodiversity of Kuči mountains. In *Katun of the Kuči Mountains*; Laković, I., Ed.; University of Montenegro: Podgorica, Montenegro, 2017; pp. 64–78.
82. Official Gazette of Montenegro, No 054/16, 15.08.2016. Law of Nature Protection of Montenegro. Available online: http://www.sluzbenilist.me/pregled-dokumenta/?id=\protect\T1\textbraceleft0C659042-9DD7-43FB-8E8C-B7DF445298A1\protect\T1\textbraceright (accessed on 23 March 2020).
83. Official Gazette of Montenegro, No 76/06, 12.12.2006, Act on Protection of Certain Plant and Animal Species. 2006. Available online: http://www.sluzbenilist.me/pregled-dokumenta/?id=\protect\T1\textbraceleft631C3E5D-4129-4985-B55B-CE4D3703CA2E\protect\T1\textbraceright (accessed on 23 March 2020).
84. Perić, B.; Baral, H.-O. *Erioscyphella curvispora*, spec. nov. from Montenegro. *Mycol. Monten.* **2015**, *17*, 89–104.
85. Baral, H.-O.; Perić, B. *Perzia triseptata* Gen. et sp. nov. (Ascomycota, Insertae sedis) from xeric bark in France and Montenegro. *Mycol. Monten.* **2016**, *19*, 7–20.
86. Baral, H.-O.; Perić, B. *Velutarina bertiscensis* and *V. alpestris* spp. nov., with a redescription of *V. rufoolivacea* and discussion of its synonyms. *Mycol. Monten.* **2015**, *17*, 17–52.
87. Perić, B.; Grebenc, T. Une espèce nouvelle du genre *Peziza* (Pezizales): *P. ontirivicola* spec. Nov. *Ascomycete. Org.* **2015**, *7*, 347–356.
88. Perić, B.; Baral, H.-O.; Partel, K. *Cenangiopsis raghavanii* and *C. junipericola* spp. nov (Cenangiaceae, Helotiales) collected in Montenegro, with redescription of a recent collection of *C. quercicola*. *Mycol. Monten.* **2015**, *17*, 7–40.
89. Perić, B.; van Vooren, N.; Healy, R.; Lazarević, J. Une *Trichophaea* rare récoltée en France et au Monténégro: *T. Flavobrunnea Comb. Nov. (Pezizales)*. *Mycol. Monten.* **2014**, *16*, 65–87.

diversity

MDPI

Article

Phylogenetic Characterization of *Botryosphaeria* Strains Associated with *Asphondylia* Galls on Species of Lamiaceae

Beata Zimowska [1], Sylwia Okoń [2], Andrea Becchimanzi [3], Ewa Dorota Krol [1] and Rosario Nicoletti [3,4,*]

1 Department of Plant Protection, University of Life Sciences, 20-068 Lublin, Poland; beata.zimowska@up.lublin.pl (B.Z.); ewa.krol@up.lublin.pl (E.D.K.)

2 Institute of Plant Genetics, Breeding and Biotechnology, University of Life Sciences, 20-068 Lublin, Poland; sylwia.okon@up.lublin.pl

3 Department of Agricultural Sciences, University of Naples Federico II, 80055 Portici, Italy; andrea.becchimanzi@unina.it

4 Council for Agricultural Research and Economics, Research Centre for Olive, Citrus and Tree Fruit, 81100 Caserta, Italy

* Correspondence: rosario.nicoletti@crea.gov.it

Received: 20 December 2019; Accepted: 20 January 2020; Published: 21 January 2020

Abstract: In the last decade, *Botryosphaeria dothidea* has been steadily reported as an associate of gall midges (Diptera, Cecidomyiidae) in a variety of host plants and ecological settings. This cosmopolitan fungus is well-known for its ability to colonize many plant species, as both a pathogen and an endophyte. Thus, the shift from this general habit to a lifestyle involving a strict symbiotic relationship with an insect introduces expectancy for possible strain specialization which could reflect separated phylogenetic lineages. Considering the recent taxonomic revision concerning species of *Botryosphaeria*, we evaluated the phylogenetic relationships among strains recovered from *Asphondylia* galls collected on several species of Lamiaceae in Poland and in Italy, and all the currently accepted species in this genus. A number of strains previously characterized from gall samples from Australia and South Africa, whose genetic marker sequences are deposited in GenBank, were also included in the analysis. As a result, full identity as *B. dothidea* is confirmed for our isolates, while strains from the southern hemisphere grouped separately, indicating the existence of genetic variation related to the geographic origin in the association with gall midges.

Keywords: *Asphondylia*; *Botryosphaeria*; *B. dothidea*; DNA sequencing; gall-associated fungi; Lamiaceae; phylogenetic relationships; symbiosis

---

1. Introduction

Although the nature of their symbiotic relationship has not been clearly ascertained, the occurrence of *Botryosphaeria dothidea* as an associate of many gall midge species (Diptera, Cecidomyiidae) is steadily reported, regardless of host plants and ecological contexts. For a long time, the identity of the fungal symbiont has been controversial, by reason of inherent difficulties in the isolation procedure, and of several taxonomic reassessments. In fact, several other fungi, such as *Cladosporium* spp. and *Alternaria* spp., have been frequently reported as gall associates, basically in connection with their saprophytic aptitude, which occasionally makes them conceal the real symbiont during the isolation procedure [1–4]. On the other hand, nomenclatural inconsistency, which only recently has been resolved after the epitypification of *B. dothidea* [5], may account for some previous incorrect reports referring to *Macrophoma*, *Diplodia*, and *Dothiorella* [2,3,6,7]. Indeed, taxonomists recommend a careful interpretation of past literature concerning this fungus [8,9].

Like several species in the Botryosphaeriaceae, *B. dothidea* is well-known for its cosmopolitan distribution and ability to colonize a high number of plants, either as a pathogen or as an endophyte [8–10]. The involvement in cecidomyid-gall formation on a variety of plants confirms it as a very adaptive species. Although the shift from association with plants to a lifestyle characterized by a strict symbiotic relationship with an insect suggests possible strain specialization, which could reflect separated phylogenetic lineages, data resulting from previous studies did not provide evidence for this hypothesis [1,11]. However, in the last decade, the evolution in fungal taxonomy boosted by the application of DNA sequencing has provided a remarkable contribution in view of a better resolution of the *Botryosphaeria* species aggregate. Six species presenting a *Fusicoccum* anamorph, namely *B. corticis*, *B. dothidea*, *B. fabicerciana*, *B. fusispora*, *B. ramosa*, and *B. scharifii*, were recognized based on a detailed phylogenetic analysis; another species, *B. agaves*, is included in this group, although its anamorphic stage has never been described so far [8]. In addition to this basic set, several new species have been more recently identified, mostly based on isolations from tree plants in China (Table 1).

Species of Lamiaceae are widespread in the Mediterranean region, where they seem to represent a diversity hotspot for gall midges. In fact, two new species of the genus *Asphondylia* have been recently described from galls collected on host plants such as *Coridothymus capitatus* [12] and *Clinopodium nepeta* [4], and two more are in course of characterization from *Micromeria graeca* and *Clinopodium vulgare* (Viggiani, personal communication). However, their distribution appears to reach Central Europe, following the geographical spread of some hosts, as documented in the case of *A. serpylli* and *A. hornigi*, respectively associated with *Thymus* spp. and *Origanum vulgare* [3,13]. In the course of our investigations on a number of species of Lamiaceae, we had the opportunity to collect isolates from galls of *Asphondylia* spp. in two European countries with different climatic conditions. Despite a certain variation in biometric characteristics, sequence homology of internal transcribed spacers of ribosomal DNA (rDNA-ITS) confirmed *B. dothidea* as the fungal symbiont of gall midges in both contexts. However, the relative unreliability of data available in GenBank for this species, which is to be taken into account after the recent rearrangements within *Botryosphaeria*, prompted us to undertake a more accurate study of the phylogenetic relationships with the new taxa and other isolates from *Asphondylia* galls from other plant species/countries, in order to assess whether or not strains adapted to this particular symbiotic relationship are homogeneous in taxonomic terms.

Table 1. *Botryosphaeria* species with *Fusicoccum* anamorph described after 2013.

Species	Main Hosts	Country	Reference
B. auasmontanum	*Acacia*	Namibia	[14]
B. kawatsukai	*Malus, Pyrus*	China	[15]
B. minutispermatia	dead wood	China	[16]
B. pseudoramosa	*Eucalyptus, Melastoma*	China	[17]
B. qinlingensis	*Quercus*	China	[18]
B. qingyuanensis	*Eucalyptus*	China	[17]
B. rosaceae	*Amygdalus, Malus, Pyrus*	China	[19]
B. sinensia	*Juglans, Malus, Morus, Populus*	China	[20]
B. wangensis	*Cedrus*	China	[17]

2. Materials and Methods

2.1. Isolation and Morphological Observations

Isolations of fungal associates of gall midges were carried out on potato dextrose agar (PDA) amended with 85% lactic acid (1 mL·L^{-1}) in 90 mm diameter Petri dishes. After removing the outer residues of the flower calyx, fragments from gall walls were cut and transferred onto the agar medium. Plates were incubated in darkness at 25 °C. Hyphal tips from the emerging fungal colonies of the botryosphaeriaceous morphotype were transferred to fresh PDA plates, for morphological observations and storage of pure cultures at 4 °C. Production of pycnidia was induced in cultures prepared in plates containing 2% water agar (WA), topped with sterilized pine needles, which were kept at room temperature under near-UV illumination [21]. Observations were carried out under a Motic BA 210 microscope (Xiamen, China), and images were taken through a 1 MP Motic camera and ScopeImage 9.0 software. Minimum, maximum, mean values, and the length/width (L/W) ratios of 50 conidia from each isolate were measured for a comparison with previously annotated reference sizes for *Botryosphaeria* species [8,17].

2.2. DNA Sequencing and Phylogenetic Analysis

Selected strains recovered from galls on several species of Lamiaceae collected in Italy and Poland (Table 2) were sampled from the surface of PDA cultures with a scalpel. The mycelial matter was transferred to 1.5 mL Eppendorf tubes, for DNA extraction. DNA isolation was performed by using a DNA easy plant and fungi isolation kit (EurX, Gdańsk, Poland), according to manufacturer's protocol. DNA concentration was estimated on 1.5% agarose gel, compared with GeneRulerTM DNA Ladder Plus (Thermo Scientific, Waltham, MA, USA), and measured by using a NanoDrop 2000 spectrophotometer (Thermo Scientific, Waltham, MA, USA). DNA samples were diluted to a concentration of 20 ng·μL^{-1} and stored at 20 °C. Amplification of loci currently considered in taxonomy of *Botryosphaeria* [8] was carried out, using primers ITS1 and ITS4 for the rDNA-ITS region [22], and primers EF1-728F and EF1-986R for the translation elongation factor 1-alpha (TEF1) region [23]. PCR reaction mixtures contained 20 ng of genomic DNA, 0.2 mM of dNTP, 0.2 mM of each primer, 1 × Taq buffer mM buffer (10 mM of Tris-HCl, 1.5 mM of $MgCl_2$, and 50 mM of KCl), and 1 U of Taq polymerase, and were adjusted to a final volume of 25 µL with sterile distilled water. PCR was conducted in a Biometra T1 thermocycler (Analytik Jena, Jena, Germany). The following reaction profile was applied: 95 °C—5 min, 35 cycles (95 °C—45 s, 52 °C—45 s, and 72 °C—45 s), with final elongation at 72 °C—5 min. PCR products were separated in 1.5% agarose gels containing EtBr in TBE buffer, at 140 V, for 1 h.

Table 2. Isolates of *B. dothidea* from *Asphondylia* galls collected on Lamiaceae used in this study.

Number	Host	Location	GenBank Accession	
			ITS	TEF1
SG3	*Clinopodium nepeta*	San Giorgio a Cremano, Italy	MN731265	MN737437
AcE3	*C. nepeta*	Astroni Nature Reserve, Italy	MN731266	MN737438
AcAs2	*C. nepeta*	Astroni Nature Reserve, Italy	MN731267	MN737439
AcSe1	*C. nepeta*	Serino, Italy	MN731268	MN737440
CLRi2	*Clinopodium vulgare*	Rivello, Italy	MN731272	MN737444
Mp26j	*Mentha piperita*	Konopnica, Poland	MN731273	MN737445
Mp30p	*M. piperita*	Konopnica, Poland	MN731274	MN737446
MgBt1	*Micromeria graeca*	Boscotrecase, Italy	MN731269	MN737441
MgPC6	*M. graeca*	Palma Campania, Italy	MN731270	MN737442
MgPC7	*M. graeca*	Palma Campania, Italy	MN731271	MN737443
OvdF3e	*Origanum vulgare*	Fajsławice, Poland	MN731275	MN737447
OvFs/g	*O. vulgare*	Fajsławice, Poland	MN731276	MN737448
Th/g2017	*Thymus vulgaris*	Fajsławice, Poland	MN731277	MN737449
ThgI/10	*T. vulgaris*	Fajsławice, Poland	MN731278	MN737450

ITS: internal transcribed spacer; TEF1: translation elongation factor 1-alpha.

After checking and determining the size of the resulting PCR products, we submitted samples to Genomed (Warsaw, Poland), for sequencing. The obtained nucleotide sequences were compared with reference strains of *Botryosphaeria* spp. from GenBank. All sequences were checked and manually edited by using CLC Main Workbench 8.1.2 software (QIAGEN, Aarhus, Denmark) where necessary. Besides our original sequences, additional sequences of *Botryosphaeria* isolates from *Asphondylia* galls were searched in GenBank for inclusion in the phylogenetic analysis, where a strain of the species *Botryobambusa fusicoccum* was used as outgroup (Table 3). The combined and single ITS and TEF1 sequences were aligned by using Muscle [24] and manually adjusted with AliView software [25], where necessary. The phylogenetic analysis was conformed to a recent protocol [26]. Congruence between the different datasets was tested by using the partition homogeneity test in PAUP software version 4.0b10 [27]. Gaps were treated as missing characters.

Phylogenetic analyses of the concatenated and single-sequence data for maximum likelihood (ML) were performed by using RAxML software version 8.2.12 [28] with GTR+G model of nucleotide substitution and 1000 bootstrap replications. Concatenated sequences were also analyzed for maximum parsimony (MP) by using PAUP, under the heuristic search parameters with tree bisection reconnection branch swapping, 100 random sequence additions, maxtrees set up to 1000, and 1000 bootstrap. Posterior probabilities of the concatenated dataset were determined by Markov Chain Monte Carlo (MCMC) sampling in MrBayes version 3.0b4 [29]. MCMC chains were run for 4000,000 generations, sampling every 100, with a 25% burn-in discarded. Phylogenetic trees were drawn by using FigTree software [30]. Both the alignments and the trees of concatenated dataset were deposited in TreeBase (http://purl.org/phylo/treebase/phylows/study/TB2:S25558).

Table 3. Reference strains used in the phylogenetic analysis.

Species	Number	GenBank Accession	
		ITS	TEF1
Botryobambusa fusicoccum	MFLUCC 11-0143	JX646792	JX646857
Botryosphaeria agaves	MFLUCC 10-0051	JX646790	JX646855
	MFLUCC 11-0125	JX646791	JX646856
Botryosphaeria auasmontanum	CBS 121769	KF766167	EU101348
	MFLUCC 15-0923	MF398858	MF398910
	MFLUCC 17-1071	MF398863	MF398915
Botryosphaeria corticis	CBS 119047	DQ299245	EU017539
	ATCC 22927	DQ299247	EU673291
Botryosphaeria dothidea	CBS 110302	AY259092	AY573218
	CBS 115476	AY236949	AY236898
	3161	EF614924	EF614940
	3179	EF614917	EF614933
	3241	EF614920	EF614937
	3242	EF614916	EF614936
	3247	EF614923	EF614941
	3253	EF614921	EF614938
	3261	EF614926	EF614943
	3275	EF614925	EF614942
	3278	EF614919	EF614934
	3279	EF614918	EF614935
Botryosphaeria fabicerciana	CMW 27094	HQ332197	HQ332213
	CMW 27108	HQ332200	HQ332216
Botryosphaeria fusispora	MFLUCC 10-0098	JX646789	JX646854
	MFLUCC 11-0507	JX646788	JX646853
Botryosphaeria kawatsukai	PGZH18	MG637267	MG637243
	PGZH19	MG637266	MG637242
Botryosphaeria minutispermatia	GZCC 16-0013	KX447675	KX447678
	GZCC 16-0014	KX447676	KX447679
Botryosphaeria pseudoramosa	CERC 2001	KX277989	KX278094
	CERC 3455	KX277997	KX278102
Botryosphaeria qinlingensis	CFCC 52984	MK434301	MK425020
	CFCC 52985	MK434302	MK425021
Botryosphaeria qingyuanensis	CERC 2946	KX278000	KX278105
	CERC 2947	KX278001	KX278106
Botryosphaeria ramosa	CBS 122069	EU144055	EU144070
Botryosphaeria rosaceae	CFCC 82350	KX197079	KX197097
	DZP B	KX197076	KX197096
Botryosphaeria scharifii	IRAN1529C	JQ772020	JQ772057
	IRAN1543C	JQ772019	JQ772056
Botryosphaeria sinensia	BJFU DZP141005-06	KT343254	KU221233
	BJFU DZP141111-10	KT343256	KU221234
Botryosphaeria wangensis	CERC 2298	KX278002	KX278107
	CERC 2300	KX278004	KX278109

3. Results

Cultures on PDA of *Botryosphaeria* isolates recovered from *Asphondylia* galls on Lamiaceae displayed a sparse to moderately dense aerial mycelium, with diverse colors, from white-cream to gray to olivaceous-black, darkening with age, occasionally with narrow or wider columns of mycelium (Figure 1A). Pycnidial conidiomata developed after 10–14 days on PDA, or 8–12 days on pine needles in WA (Figure 1B). These fruiting bodies released buff, and, respectively, black (Figure 1C) or cream masses of spores containing typical *Fusicoccum* conidia, one-celled or with one septum (Figure 1D). They were smooth, hyaline, mostly with granular content, fusiform or irregularly fusiform, wider in the middle to upper third, base-truncate or subtruncate with rounded apex, and quite variable in size (Table 4). Muriform conidia referable to the synanamorphic stage *Dichomera* were never observed, unlike what previously resulted in subcultures of strains from galls collected on *T. vulgaris* directly prepared from the isolation plates [3]. This failure was assumed to possibly derive from prolonged storage at 4 °C of the stock cultures of our strains. Likewise, no isolate produced ascomata throughout the observation period.

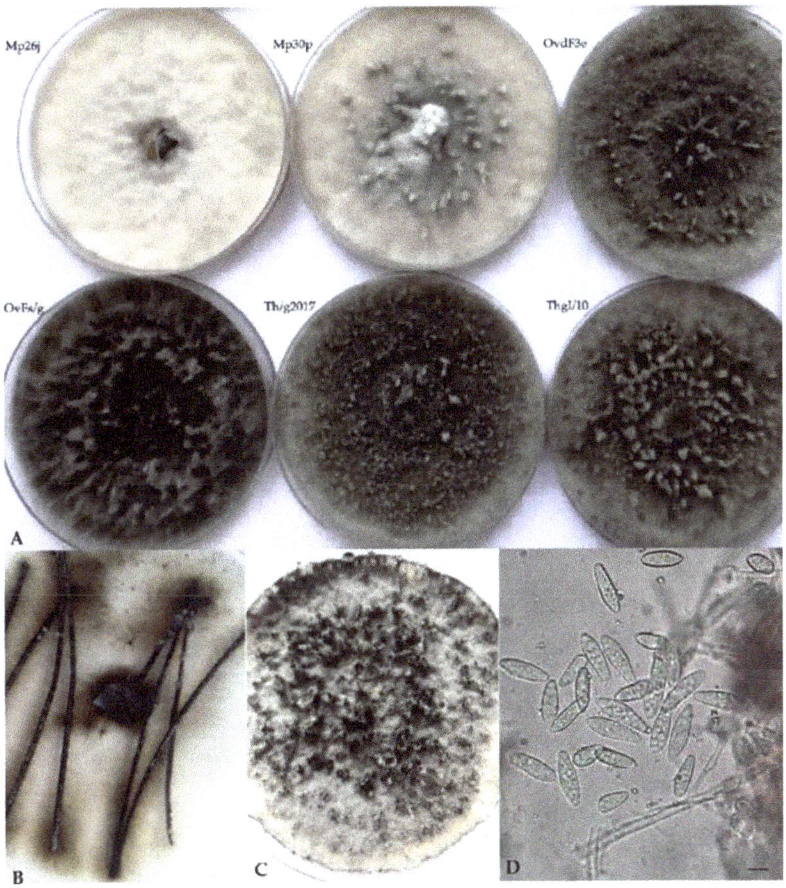

Figure 1. (**A**) Variable morphology of isolates from *Asphondylia* galls collected in our study. (**B**,**C**) Production of pycnidia, respectively, on pine needles on WA (water agar) and on PDA (potato dextrose agar). (**D**) *Fusicoccum* conidia.

Table 4. Morphological features of *B. dothidea* isolates from Lamiaceae examined in this study.

Strain Number	Conidial Dimensions (Length × Width, μm)		
	Range (50 Conidia)	Mean	Length/Width Ratio
SG3	(14–)16.8–21.8 × (3.5–)4.2–5.6	18.9 × 4.9	3.2–5.2
AcAs2	15.9–21 × (3.9–)4.3-5.8(–6.6)	18.7 × 5.2	2.6–4.7
AcE3	(14.2–)16.4–21.6 × (3.4–)4.6–5.4	19.4 × 5.1	3.1–5.4
AcSe1	(13.4–)15.3–21(–22.9) × 3.8–5.7	18.2 × 5.1	2.3–5.1
CLRi2	14.9–21.8 × (3.9–)4.3–5.8(–6.6)	18.7 × 5.2	2.6–5.4
Mp26j	(17.5–)18.5–29.7 × 3.7–7(–7.5)	22.3 × 5.1	3.5–6.1
Mp30p	(16.5–)19.0–28.7 × (3.2–)4.0–7.3	22.3 × 5.1	3.7–6.1
MgBt1	(13–)16.8-21.2 × 4.6–5.5(–6.2)	17.3 × 5.2	2.4–4.6
MgPC6	(12.8–)15.3–21 × 3.8–5.7	15.6 × 4.9	2.1–5.5
MgPC7	(13.4–)15.5–21(–22.9) × 3.3–5.7	18.9 × 4.4	2.3–6.0
OvdF3e	(17–)18.5–25.9 × (3.0–)3.5–7.4	24.3 × 5.3	3.0–6.1
OvFs/g	(17.5–)18.5–28.7 × (2.8–)3.7–7.4	23.5 × 5.1	3.5–6.0
Th/g2017	(15.2–)16.5–25.7 × (2.7–)3.9–6.8	20.3 × 4.8	2.9–5.5
ThgI/10	(16.0–)18.5–28.4 × (3.1–)3.5–7.0	22.0 × 5.1	3.4–6.0

Because of the above unreliability of morphological characters, DNA sequence homology was fundamental for an accurate taxonomic identification. PCR products of approximately 560 bp for ITS region and 300 bp for TEF1 region were amplified and successfully sequenced in 14 isolates from *Asphondylia* galls considered in this study. All the nucleotide sequences obtained were deposited in GenBank (Table 2), and blasted against the ex-epitype strain of *B. dothidea* (CMW8000/CBS115476) [5]. Identity was 99.79% for all strains but one (MgBt1, 98.97%) for ITS sequences, while for TEF1 sequence identity was 100% except four isolates (Th/g2017, OvFs/g, OvdF3e, and again MgBt1) matching at 99.16%.

This identity was highlighted in the subsequent phylogenetic analysis considering reference strains of all the recognized *Botryosphaeria* spp. producing *Fusicoccum* conidia, along with previously identified strains of *B. dothidea* collected from *Asphondylia* galls, in other contexts, worldwide. Although several contributions have been published in recent years on the subject of gall midges and associated fungi, a search in GenBank showed that sequences of both ITS and TEF1 are only available for some isolates from *Acacia* spp., collected in Australia and South Africa, which were the subject of a previously mentioned study [1].

The trimmed and manually adjusted alignment of concatenated locus contained 58 strains (including the outgroup) and consisted of 490 and 255 bp for ITS and TEF1, respectively. The best scoring RAxML tree (Figure 2) had a final likelihood value of −1961.235515. The matrix had 172 distinct alignment patterns, with 5.74% of undetermined characters or gaps. The ML tree of ITS alone showed poor resolution compared to ML trees based on TEF1 and concatenated sequences (Figure S1). ML trees of TEF1 and ITS + TEF1 showed almost the same topology, except for *B. qingyuanensis*, which is included in the same clade of *B. wangensis*-*B. sinensis*-*B. qinlingensis* in the phylogram based on TEF1 alone (Figure S2). Parsimony analysis yielded 1000 equally parsimonious trees (tree length = 174 steps; consistency index = 0.902; retention index = 0.915; relative consistency index = 0.826; homoplasy index = 0.098). Of the 745 characters used, 76 were parsimony-informative, 65 were variable and parsimony-uninformative, and 604 were constant. The same clades were supported in MP and ML analyses, except for *B. minutispermatia*, which resulted in being less closely related to the Australian isolates in the MP tree (Figure S3).

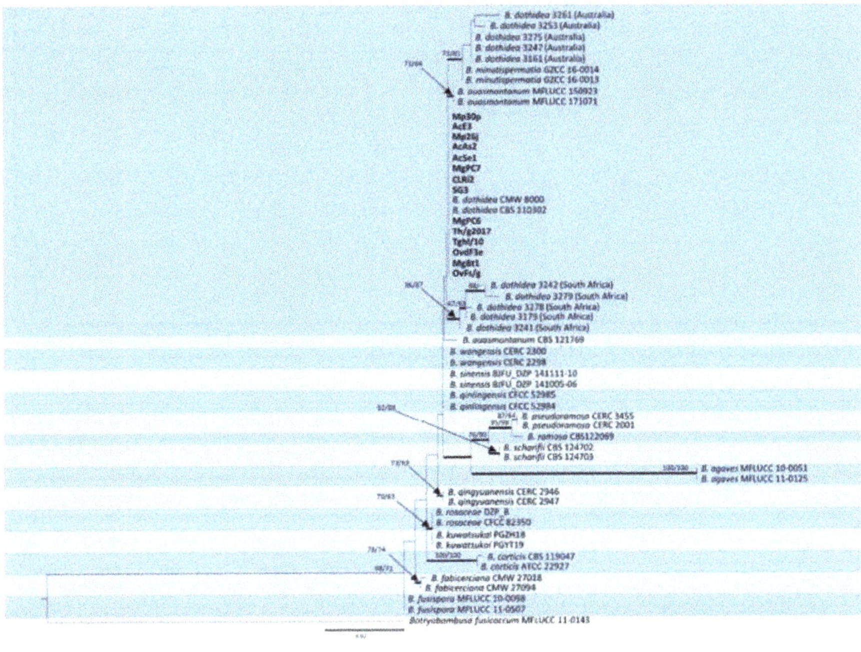

Figure 2. Phylogenetic tree based on maximum likelihood (ML) analyses of concatenated ITS and TEF1 sequences from strains considered in this study. Original isolates from Lamiaceae are in bold. Bootstrap support values ≥ 60% for ML and maximum parsimony (MP) are presented above branches as follows: ML/MP, bootstrap support values < 60% are marked with '-'. Branches in bold are supported by Bayesian analysis (posterior probability ≥ 90%). *Botryobambusa fusicoccum* MFLUCC 11-0143 was used as the outgroup reference.

4. Discussion

The wide morphological variation observed in our study is in line with recent findings that biometric data of *Botryosphaeria* species may overlap and are no longer relevant for the identification of *B. dothidea* [8,9,17]. Even *Dichomera* conidia are infrequently observed in this species [8,14,21], indicating that this character cannot be reliably taken into account for taxonomic purposes. Hence, reports from *Asphondylia* galls primarily referring to this anamorphic stage [1] should be considered with caution, because of the common co-occurrence of saprophytic *Alternaria* producing similar phaeodictyospores.

Therefore, nucleotide sequencing has become the primary identification method for *B. dothidea* through the assessment of homologies with the ex-epitype strain (CMW8000/CBS115476) [5], particularly considering the loci of ITS and TEF1. In fact, although ITS was shown to clearly distinguish *B. dothidea* from its closest relatives, the recent discovery of several cryptic species within the Botryosphaeriaceae makes it necessary to combine with TEF1 for a more reliable identification [8,31,32].

Phylogenetic analysis disclosed a clear identity with *B. dothidea* of our heterogeneous strain sample from Lamiaceae, regardless of their origin from two different climatic contexts. In fact, both Polish and Italian isolates grouped together with the type strains of this species. Conversely, Australian and South African strains formed two distinct groupings, which indicated a divergence from the most recent common ancestor. Particularly, the former was associated to type strains of the species *B. minutispermatia* in a more comprehensive clade, while the latter exhibited a slightly higher phylogenetic distance. Considering that both Australian and South African isolates were collected in association with gall midges on *Acacia* spp., this finding could be interpreted as being in agreement with a previous analysis of the bulk *B. dothidea* sequences deposited in GenBank, showing a population structure which is shaped by geographical distance rather than host-plant preference [9]. In fact, lower identity with *B. dothidea* strains in the GenBank database resulted for these isolates, particularly those from South Africa whose TEF1 sequence homology was not higher than 96.83%. The observed variation particularly concerning this genetic marker requires further assessments to establish if these clusters should be interpreted as separated lineages within *B. dothidea*, or if they may represent distinct species.

The case of *B. auasmontanum*, a species characterized from a single strain (CBS121769) collected in Namibia [14], deserves further consideration. In fact, besides the holotype, our analysis included two more isolates from Italy ascribed to this species [26], whose ITS and TEF1 sequences are available in GenBank. Unexpectedly, they did not cluster with the holotype in the phylogenetic tree (Figure 2). A BLAST search in GenBank with sequences by these two strains clearly shows their 100% identity with *B. dothidea*, while ITS and TEF1 sequence homology with CBS121769 is lower (95.21% and 90.28%, respectively). The phylogenetic separation of the holotype of *B. auasmontanum* is supported by two large gaps resulting from the sequence alignment (Figure 3), which means that the claimed evidence of possible synonymy between the two species [26] applies only to the two Italian isolates. Thus, their inclusion in phylogenetic studies as representatives of *B. auasmontanum* should be avoided, and their incorrect taxonomic identification should be taken into account. Figure 3 also describes the substantial similarity among all strains of *B. dothidea*, including Australian and South African isolates, and the closely related *B. minutispermatia*, which essentially differ for nucleotide substitutions at single definite positions.

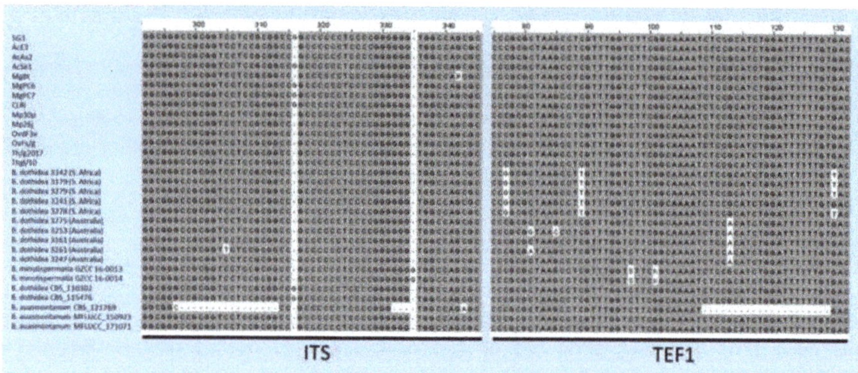

Figure 3. Alignment of ITS and TEF1 sequences of strains of *B. dothidea*, *B. minutispermatia*, and *B. auasmontanum*. Gaps in the sequences of the holotype of the latter species (CBS121769) are evident, along with single nucleotide substitutions in TEF1 sequences of Australian and South African isolates, and the type strains of *B. minutispermatia*.

5. Conclusions

Characterization of *Botryosphaeria* strains from different plant species and environmental contexts goes through the finding of novel species, with the expectation that additional undescribed taxa may be discovered in the future [17]. This concept involves strains associated with cecidomyids, inducing gall formation on a wide variety of plant species, which so far have been overlooked in assessments concerning taxonomy of such a highly adaptive and widespread fungus. Results of our investigation showed full identity with *B. dothidea* of isolates from galls collected from Lamiaceae, while a possible separation from this species should be verified for isolates previously recovered from *Acacia* in Australia and, particularly, South Africa. Indeed, a more adequate definition could be obtained by integrating these findings with data concerning cecidomyid-associated *Botryosphaeria* strains from other countries.

Therefore, a more active cooperation among researchers currently working on this topic worldwide is to be encouraged, in order to shed further light on this unique biological association.

Supplementary Materials: The following are available online at http://www.mdpi.com/1424-2818/12/2/41/s1. Figure S1: Maximum-likelihood tree of ITS sequence. Figure S2: Maximum-likelihood tree of EF sequence. Figure S3: One of the 1000 most parsimonious trees resulting from the analysis of concatenated ITS and TEF1 sequences.

Author Contributions: Conceptualization, B.Z. and R.N.; methodology, A.B., B.Z., E.D.K., and S.O.; formal analysis, A.B. and S.O.; writing—original draft preparation, B.Z., R.N., and S.O.; writing—review and editing, B.Z., E.D.K., and R.N.; funding acquisition, E.D.K. All authors have read and agreed to the published version of the manuscript.

Funding: This research received no external funding.

Acknowledgments: Authors thank Sarah Lucchesi (University of Southern Maine, Portland, USA) for revising the English style of this paper.

Conflicts of Interest: The authors declare no conflict of interest.

References

1. Adair, R.J.; Burgess, T.; Serdani, M.; Barber, P. Fungal associations in *Asphondylia* (Diptera: Cecidomyiidae) galls from Australia and South Africa: Implications for biological control of invasive acacias. *Fungal Ecol.* **2009**, *2*, 121–134. [CrossRef]
2. Lebel, T.; Peele, C.; Veenstra, A. Fungi associated with *Asphondylia* (Diptera: Cecidomyiidae) galls on *Sarcocornia quinqueflora* and *Tecticornia arbuscula* (Chenopodiaceae). *Fungal Divers.* **2012**, *55*, 143–154. [CrossRef]
3. Zimowska, B.; Viggiani, G.; Nicoletti, R.; Furmańczyk, A.; Becchimanzi, A.; Kot, I. First report of the gall midge *Asphondylia serpylli* on thyme (*Thymus vulgaris*), and identification of the associated fungal symbiont. *Ann. Appl. Biol.* **2017**, *171*, 89–94. [CrossRef]
4. Bernardo, U.; Nugnes, F.; Gualtieri, L.; Nicoletti, R.; Varricchio, P.; Sasso, R.; Viggiani, G. A new gall midge species of *Asphondylia* (Diptera: Cecidomyiidae) inducing flower galls on *Clinopodium nepeta* (Lamiaceae) from Europe, its phenology, and associated fungi. *Environ. Entomol.* **2018**, *47*, 609–622. [CrossRef] [PubMed]
5. Slippers, B.; Crous, P.W.; Denman, S.; Coutinho, T.A.; Wingfield, B.D.; Wingfield, M.J. Combined multiple gene genealogies and phenotypic characters differentiate several species previously identified as *Botryosphaeria dothidea*. *Mycologia* **2004**, *96*, 83–101. [CrossRef]
6. Denman, S.; Crous, P.W.; Taylor, J.E.; Kang, J.-C.; Pascoe, I.; Wingfield, M.J. An overview of the taxonomic history of *Botryosphaeria*, and a re-evaluation of its anamorphs based on morphology and ITS rDNA phylogeny. *Stud. Mycol.* **2000**, *45*, 129–140.
7. Park, I.; Sanogo, S.; Hanson, S.F.; Thompson, D.C. Molecular identification of *Botryosphaeria dothidea* as a fungal associate of the gall midge *Asphondylia prosopidis* on mesquite in the United States. *BioControl* **2019**, *64*, 209–219. [CrossRef]
8. Phillips, A.J.L.; Alves, A.; Abdollahzadeh, J.; Slippers, B.; Wingfield, M.J.; Groenewald, J.Z.; Crous, P.W. The Botryosphaeriaceae: genera and species known from culture. *Stud. Mycol.* **2013**, *76*, 51–167. [CrossRef]
9. Marsberg, A.; Kemler, M.; Jami, F.; Nagel, J.H.; Postma-Smidt, A.; Naidoo, S.; Wingfield, M.J.; Crous, P.W.; Spatafora, J.W.; Hesse, C.N.; et al. *Botryosphaeria dothidea*: a latent pathogen of global importance to woody plant health. *Mol. Plant Pathol.* **2017**, *18*, 477–488. [CrossRef]
10. Slippers, B.; Wingfield, M.J. Botryosphaeriaceae as endophytes and latent pathogens of woody plants: diversity, ecology and impact. *Fungal Biol. Rev.* **2007**, *21*, 90–106. [CrossRef]
11. Janson, E.M.; Peeden, E.R.; Stireman, J.O.; Abbot, P. Symbiont-mediated phenotypic variation without co-evolution in an insect–fungus association. *J. Evol. Biol.* **2010**, *23*, 2212–2228. [CrossRef] [PubMed]
12. Malagaris, P. Biology and ecology of *Asphondylia coridothymi* (Diptera: Cecidomyiidae) inducing galls on *Coridothymus capitatus* on the island of Samos, Greece. *Acta Soc. Zool. Bohemicae* **2011**, *75*, 239–251.
13. Viggiani, G.; Stinca, A. Infestations on oregano by the gall midge *Asphondylia hornigi* Wachtl. *Prot. Colture* **2018**, *4*, 29–31.
14. Slippers, B.; Roux, J.; Wingfield, M.J.; van der Walt, F.J.J.; Jami, F.; Mehl, J.W.M.; Marais, G.J. Confronting the constraints of morphological taxonomy in the Botryosphaeriales. *Persoonia* **2014**, *33*, 155–168. [CrossRef]

15. Xu, C.; Wang, C.S.; Ju, L.L.; Zhang, R.; Biggs, A.R.; Tanaka, E.; Li, B.; Sun, G.Y. Multiple locus genealogies and phenotypic characters reappraise the causal agent of apple ring rot in China. *Fungal Divers.* **2015**, *71*, 215–231. [CrossRef]
16. Ariyawansa, H.A.; Hyde, K.D.; Liu, J.K.; Wu, S.P.; Liu, Z.Y. Additions to Karst Fungi 1: *Botryosphaeria minutispermatia* sp. nov., from Guizhou Province, China. *Phytotaxa* **2016**, *275*, 35–44. [CrossRef]
17. Li, G.Q.; Liu, F.F.; Li, J.Q.; Liu, Q.L.; Chen, S.F. Botryosphaeriaceae from *Eucalyptus* plantations and adjacent plants in China. *Persoonia* **2018**, *40*, 63–95. [CrossRef]
18. Liang, L.Y.; Jiang, N.; Chen, W.Y.; Liang, Y.M.; Tian, C.M. *Botryosphaeria qinlingensis* sp. nov. causing oak frogeye leaf spot in China. *Mycotaxon* **2019**, *134*, 463–473. [CrossRef]
19. Zhou, Y.P.; Zhang, M.; Dou, Z.P.; Zhang, Y. *Botryosphaeria rosaceae* sp. nov. and *B. ramosa*, new botryosphaeriaceous taxa from China. *Mycosphere* **2017**, *8*, 162–171. [CrossRef]
20. Zhou, Y.; Dou, Z.; He, W.; Zhang, X.; Zhang, Y. *Botryosphaeria sinensia* sp nov., a new species from China. *Phytotaxa* **2016**, *245*, 43–50. [CrossRef]
21. Crous, P.W.; Slippers, B.; Wingfield, M.J.; Rheeder, J.; Marasas, W.F.; Philips, A.J.; Alves, A.; Burgess, T.; Barber, P.; Groenewald, J.Z. Phylogenetic lineages in the Botryosphaeriaceae. *Stud. Mycol.* **2006**, *55*, 235–253. [CrossRef] [PubMed]
22. White, T.J.; Bruns, T.; Lee, S.J.W.T.; Taylor, J. Amplification and direct sequencing of fungal ribosomal RNA genes for phylogenetic. In *PCR Protocols, A Guide to Methods and Applications*; Innis, M.A., Gelfand, D.H., Sninsky, J.J., White, T.J., Eds.; Academic Press: New York, NY, USA, 1990; Volume 18, pp. 315–322.
23. Carbone, I.; Kohn, L.M. A method for designing primer sets for speciation studies in filamentous ascomycetes. *Mycologia* **1999**, *91*, 553–556. [CrossRef]
24. Edgar, R.C. MUSCLE: Multiple sequence alignment with high accuracy and high throughput. *Nucleic Acids Res.* **2004**, *32*, 1792–1797. [CrossRef] [PubMed]
25. Larsson, A. AliView: A fast and lightweight alignment viewer and editor for large datasets. *Bioinformatics* **2014**, *30*, 3276–3278. [CrossRef]
26. Dissanayake, A.J.; Camporesi, E.; Hyde, K.D.; Yan, J.Y.; Li, X.H. Saprobic Botryosphaeriaceae, including *Dothiorella italica* sp. nov., associated with urban and forest trees in Italy. *Mycosphere* **2017**, *8*, 1157–1176. [CrossRef]
27. Wilgenbusch, J.C.; Swofford, D. Inferring evolutionary trees with PAUP*. *Curr. Prot. Bioinformatics* **2003**, *1*, 6.4.1–6.4.28. [CrossRef]
28. Stamatakis, A. RAxML-VI-HPC: maximum likelihood-based phylogenetic analyses with thousands of taxa and mixed models. *Bioinformatics* **2006**, *22*, 2688–2690. [CrossRef]
29. Huelsenbeck, J.P.; Ronquist, F. MRBAYES: Bayesian inference of phylogenetic trees. *Bioinformatics* **2001**, *17*, 754–755. [CrossRef]
30. Molecular Evolution, Phylogenetics and Epidemiology. Available online: http://tree.bio.ed.ac.uk/software/figtree/ (accessed on 25 November 2018).
31. Chen, S.; Pavlic, D.; Roux, J.; Slippers, B.; Xie, Y.; Wingfield, M.J.; Zhou, X.D. Characterization of Botryosphaeriaceae from plantation-grown *Eucalyptus* species in South China. *Plant Pathol.* **2011**, *60*, 739–751. [CrossRef]
32. Abdollahzadeh, J.; Zare, R.; Phillips, A.J. Phylogeny and taxonomy of *Botryosphaeria* and *Neofusicoccum* species in Iran, with description of *Botryosphaeria scharifii* sp. nov. *Mycologia* **2013**, *105*, 210–220. [CrossRef]

diversity

MDPI

Article

News from the Sea: A New Genus and Seven New Species in the Pleosporalean Families Roussoellaceae and Thyridariaceae

Anna Poli, Elena Bovio, Lucrezia Ranieri, Giovanna Cristina Varese * and Valeria Prigione

Mycotheca Universitatis Taurinensis, Department of Life Sciences and Systems Biology, University of Torino, Viale Mattioli 25, 10125 Torino, Italy; anna.poli@unito.it (A.P.); elena.bovio@inrae.fr (E.B.); lucrezia.ranieri@edu.unito.it (L.R.); valeria.prigione@unito.it (V.P.)

* Correspondence: cristina.varese@unito.it; Tel.: +39-011-670-5964

Received: 19 March 2020; Accepted: 3 April 2020; Published: 6 April 2020

Abstract: Nineteen fungal strains associated with the seagrass *Posidonia oceanica*, with the green alga *Flabellia petiolata*, and the brown alga *Padina pavonica* were collected in the Mediterranean Sea. These strains were previously identified at the family level and hypothesised to be undescribed species. Strains were examined by deep multi-loci phylogenetic and morphological analyses. Maximum-likelihood and Bayesian phylogenies proved that *Parathyridariella* gen. nov. is a distinct genus in the family Thyriadriaceae. Analyses based on five genetic markers revealed seven new species: *Neoroussoella lignicola* sp. nov., *Roussoella margidorensis* sp. nov., *R. mediterranea* sp. nov., and *R. padinae* sp. nov. within the family Roussellaceae, and *Parathyridaria flabelliae* sp. nov., *P. tyrrhenica* sp. nov., and *Parathyridariella dematiacea* gen. nov. et sp. nov. within the family Thyridariaceae.

Keywords: marine fungi; new taxa; phylogeny; lignicolous fungi

1. Introduction

Marine fungi are a relevant and active component of the microbial communities that inhabit the oceans [1]. Fungi in the marine environment live as mutualists, parasites, pathogens and saprobes, and are pivotal to marine food webs because of the recycling of recalcitrant substrata [2]; besides, these widely dispersed organisms are a source of novel bioactive compounds [3].

Marine fungi have been recovered worldwide from a broad range of biotic and abiotic substrata, such as driftwood algae, sponges, corals, sediments, etc. [4,5]. Following the definition of Pang et al [6] that considered "a marine fungus" to be any fungus retrieved repeatedly from marine environment and that reproduces in the marine environment, Jones et al. [7] listed 1680 fungal species belonging to 693 genera, 223 families, 87 orders, 21 classes and six phyla. However, considering that the total number of marine fungi has been estimated to exceed 10,000 taxa [8], fungal diversity remains largely undescribed. With more than 900 species [9], the Ascomycota are the dominant fungal phylum in the sea; the most represented lineages include the order Pleosporales (class Dothideomycetes) with 36 families, 95 genera and 194 species described to date (www.marinefungi.org).

In recent surveys aimed to uncover the underwater fungal diversity, 19 unidentified Roussoellaceae were isolated from several substrates, as follows: 12 from the brown alga *Padina pavonica* (L.) Thivy [10], 4 from the green alga *Flabellia petiolata* (Turra) Nizamuddin [11], 2 from the seagrass *Posidonia oceanica* (L.) [12] Delile, and 1 from the Atlantic sponge *Dysidea fragilis* (Montagu) [13]. The Roussoellaceae is a well-resolved family in the Pleosporales [14]. Others [15] have treated the family Roussoellaceae as a synonym of Thyridariaceae, based on phylogenetic affinities. However, following the discovery of new genera in this group, delineated by high resolution multi-locus phylogenetic analyses, the Roussoellaceae and Thyridariaceae are now recognized as two distinct but closely related families [16–20].

Many new species of Roussoellaceae and Thyridariaceae have recently been described on terrestrial plants including bamboo, palms and mangroves [14,17,20,21]. This paper provides a more precise phylogenetic placement of the 19 strains isolated from marine substrata together with morphological insights of those strains that represent new species within these two families.

2. Materials and Methods

2.1. Fungal Isolates

The fungal isolates analyzed in this paper were retrieved in the Mediterranean Sea from *P. oceanica* (2), collected in Riva Trigoso bay and Elba island, *P. pavonica* (12), and *F. petiolata* (3) from the coastal waters of Elba island [10–12]. A single isolate was previously retrieved in association with *D. fragilis* in the Atlantic Ocean [13] (Table 1).

Table 1. Dataset used for phylogenetic analysis. Genbank sequences including newly generated nrITS, nrLSU, nrSSU, TEF1-α and RPB2 amplicons relative to the novel species of Roussoellaceae and Thyridariaceae, to *Parathyridaria robiniae* MUT 2452 and MUT 4893 and to *Parathyridaria ramulicola* MUT 4397.

Species	Strain Code	Source	nrITS	nrSSU	nrLSU	TEF-1α	RPB2
Roussoellaceae							
***Arthopyrenia salicis* Massal**	CBS 368.94	*Salix bark*	KF443410	AY538333	AY538339	KF443404	KF443397
***Neoroussoella bambusae* Liu and Hyde**	MFLUCC 11-0124	Dead branch of *Bambusa*	KJ474827	–	KJ474839	KJ474848	KJ474856
***N. alishanense* Karunarathna, Kuo, Phookamsak and Hyde**	AKTW 03 FU31016	*Pennisetum purpureum*	MK503816	MK503828	MK503822	MK336181	MN037756
	AKTW 11 FU31018	*Pennisetum purpureum*	MK503818	MK503830	MK503824	MK336182	MN037757
***N. entadae* Jones and Hyde**	MFLUCC 18-0243	*Leucaena sp.*	MK347786	MK347893	MK348004	MK360065	MK434866
***N. heveae* Senwanna, Phookamsak and Hyde**	MFLUCC 17-1983	Twig of *Hevea brasiliensis*	MH590693	–	MH590689	–	–
***N. leucaenae* Jones and Hyde**	MFLUCC 18-1544	Decaying pod of *Leucaena*	MK347767	MK347874	MK347984	MK360067	MK434876
	MFLUCC 17-0927	*Pterocarpus* sp.	MK347733	MK347841	MK347950	MK360066	MK434896
***Neoroussoella lignicola*sp. nov.**	**MUT 4904**	*P. pavonica*	KT699129	**MN556307***	**MN556319***	**MN605894***	**MN605914***
	MUT 5008	*P. oceanica* leaves	**MN556317***	**MN556308***	**MN556320***	**MN605895***	**MN605915***
	MUT 5373	*P. pavonica*	KU314953	KU314954	**MN556321***	**MN605896***	**MN605916***
***Pararoussoella juglandicola* Crous and Schumach**	CBS 145037		MK442607	–	MK442543	MK442699	MK442671
***P. mukdahanensis* (Phookamsak, Dai and Hyde) Crous**	MFLUCC 11-0201	Bamboo	KU940129	KU872121	KU863118	–	–
***P. rosarum* Jones and Hyde**	MFLUCC 17-0796	*Rosa* sp.	MG8289391	MG829154	MG829048	MG829224	–
***Pseudoneoconiothyrium rosae* (Phukhams., Camporesi and Hyde) Phukhams., Camporesi and Hyde**	MFLUCC 15-0052	Dead aerial spines of *Rosa canina*	MG828922	MG829138	MG829032	–	–

Table 1. *Cont.*

Species	Strain Code	Source	nrITS	nrSSU	nrLSU	TEF-1α	RPB2
Roussoella chiangraina **Phookamsak, Liu and Hyde**	MFLUCC 10-0556	Dead branch of bamboo	KJ474828	–	KJ474840	KJ474849	KJ474857
R. doimaesalongensis **Thambug. and Hyde**	MFLUCC 14–0584	Dead branch of bamboo	KY026584	–	KY000659	KY651249	KY678394
R. elaeicola **Konta. and Hyde**	MFLUCC 15-0276a	Dead petiole of *Elaeis guineensis*	MH742329	–	MH742326	–	–
	MFLUCC 15-0276b	Dead petiole of *Elaeis guineensis*	MH742330	–	MH742327	–	–
R. euonymi **Crous and Akulov**	CBS 143426	Fallen branches of *Euonymus europaeus*	MH107915	–	MH107961	–	MH108007
R. hysterioides **(Ces.) Höhn.**	CBS 546.94	*Phyllostachys*	KF443405	AY642528	KF443381	KF443399	KF443392
R. intermedia **Ju, Rogers and Huhndorf**	CBS 170.96	Bamboo	KF443407	KF443390	KF443382	KF443398	KF443394
R. japanensis**Kaz. Tanaka, Liu and Hyde**	MAFF 239636	Twigs of *Sasa veitchii*	KJ474829	AB524480	AB524621	AB539114	AB539101
R. kunmingensis **Jiang, Phookamsak and Hyde**	KUMCC 18-0128	Dead bamboo	MH453491	–	MH453487	MH453480	MH453484
R. mangrovei **Phukhams. and Hyde**	MFLUCC 16-0424	Dead branches of *Rhizophora*	MH025951	–	MH023318	MH028246	MH028250
Roussoella margidorensis **sp. nov.**	**MUT 5329**	*P. pavonica*	KU314944	**MN556309***	**MN556322***	**MN605897***	**MN605917***
Roussoella mediterranea **sp. nov.**	**MUT 5306**	*P. pavonica*	KU255054	**MN556310***	**MN556323***	**MN605898***	**MN605918***
	MUT 5369	*P. pavonica*	KU314947	KU314948	**MN556324***	**MN605899***	**MN605919***
R. mexicana **Crous and Yáñez-Mor.**	CPC 25355	Leaf spots of Coffea arabica	KT950848	–	KT950862	–	–
R. neopustulans **Dai, Liu and Hyde**	MFLUCC 11-0609	Bamboo	KJ474833	–	KJ474841	KJ474850	
	MFLUCC 12-0003	Bamboo	KU940130	KU872122	KU863119	–	–
R. nitidula **Sacc. and Paol.**	MFLUCC 11-0182	Bamboo	KJ474835	–	KJ474843	KJ474852	KJ474859
	MFLUCC 11-0634	Bamboo	KJ474834	–	KJ474842	KJ474851	KJ474858
Roussoella padinae **sp. nov.**	**MUT 5341**	*P. pavonica*	KU158153	KU158176	**MN556325***	**MN605900***	**MN605920***
	MUT 5365	*P. pavonica*	KU158170	KU158179	**MN556326***	**MN605901***	**MN605921***
	MUT 5503	*P. pavonica*	KU158170	**MN556312***	**MN556327***	**MN605902***	**MN605922***
R. pseudohysterioides **Dai and Hyde**	MFLUCC 13-0852	Bamboo	KU940131	KU872123	KU863120	KU940198	–
R. pustulans **(Ellis and Everh.) Ju, Rogers and Huhndorf**	KT 1709	Culms of *Sasa kurilensis*	KJ474830	AB524482	AB524623	AB539116	AB539103
R. scabrispora **(Höhn.) Aptroot**	MFLUCC 11-0624	Bamboo	KJ474836	–	KJ474844	KJ474853	KJ474860
	RSC	Bamboo	KX650566	–	KX650566	KX650537	–
R. siamensis **Phookamsak, Liu and Hyde**	MFLUCC 11-0149	Bamboo	KJ474837	KU872125	KJ474845	KJ474854	KJ474861
R. thailandica **Dai, Liu and Hyde**	MFLUCC 11-0621	Bamboo	KJ474838	–	KJ474846	–	
R. tuberculata **Dai and Hyde**	MFLUCC 13-0854	Bamboo	KU940132	KU872124	KU863121	KU940199	
R. verrucispora **Kaz. Tanaka, Liu and Hyde**	CBS 125434	*Sasa kurilensis*	KJ474832	AB52448	AB524622	AB539115	–

Table 1. *Cont.*

Species	Strain Code	Source	nrITS	nrSSU	nrLSU	TEF-1α	RPB2
R. yunnanensis **Jiang, Phookamsak and Hyde**	KUMCC 18-0115	Dead bamboo	MH453492	–	MH453488	MH453481	–
Roussoellopsis macrospora **(Hino and Katum.) Hinoand Katum**	MFLUCC 12-0005	Bamboo	KJ739604	KJ739608	KJ474847	KJ474855	KJ474862
Ro. tosaensis **(Hino and Katum.) Hino and Katum**	KT 1659	Culms of bamboo	–	AB524484	AB524625	AB539117	AB539104
Thyridariaceae							
Cycasicola goaensis **Jones and Hyde**	MFLUCC 17-0754	*Cycas* sp.	MG828885	MG829112	MG829001	MG829198	–
C. leucaenae **Jones and Hyde**	MFLUCC 17-0914	Leucaena leucocephala	MK34772	MK347833	MK347942	MK360046	MK434900
Liua muriformis **Phookamsak, Jiang and Hyde**	KUMCC 18-0177	Dead hanging branches of *Lonicera maackii*	MK433599	MK433595	MK433598	MK426798	MK426799
Parathyridaria percutanea **(Ahmed, Stevens, van de Sande and de Hoog) Jaklitsch and Voglmayr**	CBS 868.95	Human	KF322118	KF366451	KF366449	KF407987	KF366452
	CBS 128203	Human	KF322117	KF366450	KF366448	KF407988	KF366453
P. ramulicola **Jaklitsch, Fourn and Voglmayr**	CBS 141479	Twigs of *Ribes rubrum*	NR_147657	KX650514	KX650565	KX650536	KX650584
	MUT 4397	*P. oceanica*	KC339235	**MN556311***	KF636775	**MN605913***	**MN605933***
P. robiniae**Mapook, Camporesi and Hyde**	MFLUCC 14-1119	Dead branch of *Robinia pseudoacacia*	KY511142	–	KY511141	KY549682	–
	MUT 2452	*Dysidea fragilis*	MG813183	**MN556312***	MG816491	**MN605903***	**MN605923***
	MUT 4893	*P. pavonica*	KM355998	KM355993	**MN556328***	**MN605904***	**MN605924***
Parathyridaria flabelliae **sp. nov.**	**MUT 4859**	*F. petiolata*	KR014355	KT587315	KP671716	**MN605909***	**MN605929***
	MUT 4886	*F. petiolata*	KR014358	KT587317	KP671720	**MN605910***	**MN605930***
Parathyridaria tyrrhenica **sp. nov.**	**MUT 4966**	*F. petiolata,*	KR014366	KT587309	KP671740	**MN605911***	**MN605931***
	MUT 5371	*P. pavonica*	KU314951	KU314952	**MN556329***	**MN605912***	**MN605932***
Parathyridariella dematiacea **sp. nov.**	**MUT 4419**	*P. oceanica* rhizomes	KC339245	**MN556313***	KF636786	**MN605905***	**MN605925***
	MUT 4884	*F. petiolata*	**MN556317***	KT587329	KP671726	**MN605906***	**MN605926***
	MUT 5310	*P. pavonica*	KU255057	**MN556314***	**MN556330***	**MN605907***	**MN605927***
	MUT 5381	*P. pavonica*	KU314959	KU314960	**MN556331***	**MN605908***	**MN605928***
Thyridaria acaciae **(Crous and Wingf.) Jaklitsch and Voglmayr**	CBS 138873	Leaves of *Acacia tortilis*	KP004469	–	KP004497	–	–
T. broussonetiae **(Sacc.) Traverso**	TB	*Hippocrepis emerus*	KX650567	–	KX650567	KX650538	KX650585
	TB1	*Amorpha fruticosa*	KX650568	KX650515	KX650568	KX650539	KX650586
Thyridariella mahakoshae **Devadatha, Sarma, Wanas., Hyde and Jones**	NFCCI 4215	Decaying wood *Avicennia marina*	MG020435	MG020441	MG020438	MG023140	MG020446
Th. mangrovei **Devadatha, Sarma, Hyde, Wanas. and Jones**	NFCCI 4213	Decaying wood *Avicennia marina*	MG020434	MG020440	MG020437	MG020443	MG020445
	NFCCI 4214	Decaying wood *Avicennia marina*	MG020436	MG020442	MG020439	MG020444	MG020447

Table 1. *Cont.*

Species	Strain Code	Source	nrITS	nrSSU	nrLSU	TEF-1α	RPB2
Occultibambusaceae							
***Occultibambusa bambusae* Dai and Hyde**	MFLUCC 11-0394	Bamboo	KU940124	–	KU863113	KU940194	KU940171
	MFLUCC 13-0855	Bamboo	KU940123	KU872116	KU863112	KU940193	KU940170
Ohleriaceae							
***Ohleria modesta* Fuckel**	MGC		KX650562	–	KX650562	KX650533	KX650582
	OM	Branches of *Chamaecytisus proliferus*	KX650563	KX650513	KX650563	KX650534	KX650583
Torulaceae							
***Dendryphion europaeum* Crous and Schumacher**	CPC 22943	*Heracleum sphondylium*	KJ869146	–	KJ869203	–	–
***Torula herbarum* (Pers.) Link**	CBS 111855	n.a.	KF443409	KF443391	KF443386	KF443403	KF443396
	CBS 595.96		KF443408	KF443387	KF443385	KF443402	KF443395
***Torula hollandica* Crous**	CBS 220.69	Delphinium dead stem	KF443406	–	MH877717	–	–

* = newly generated sequences; n.a. = not available.

The strains investigated were originally isolated on Corn Meal Agar medium supplemented with sea salts (CMASS; 3.5% w/v sea salt mix, Sigma-Aldrich, Saint Louis, USA, in ddH_2O) and are preserved at the *Mycotheca Universitatis Taurinensis* (MUT), Italy.

2.2. Morphological Analysis

All isolates were pre-grown on Malt Extract Agar-sea water (MEASW; 20 g malt extract, 20 g glucose, 2 g peptone, 20 g agar in 1 L of sea water) for one month at 24 °C prior to inoculation in triplicate onto new Petri dishes (9 cm Ø) containing (i) MEASW, (ii) Oatmeal Agar-sea water (OASW; 30 g oatmeal, 20 g agar in 1 L of sea water), or iii) Potato Dextrose Agar-sea water (PDASW; 4 g potato extract, 20 g dextrose, 20 g agar in 1 L of sea water). Petri dishes were incubated at 15 and/or 24 °C. The colony growth was monitored periodically for 28 days. Macroscopic and microscopic traits, were assessed for strains grown on MEASW at the end of the incubation period.

In an attempt to induce sporulation, sterile pieces of *Quercus ruber* cork and *Pinus pinaster* wood (species autochthonous to the Mediterranean area) were placed on 3 week old fungal colonies grown on MEASW ([22], modified). Petri dishes were further incubated for 4 weeks at 24 °C. Subsequently, cork and wood pieces were transferred to 50 mL tubes containing 20 mL of sterile sea water. Samples were incubated at 24 °C for one month. In parallel, the strains were also plated on Syntetic Nutrient Agar-sea water (SNASW; 1 g KH_2PO_4, 1 g KNO_3, 0.5 g $MgSO_4 \bullet 7H_2O$, 0.5 g KCl, 0.2 g glucose, 0.2 g sucrose, 20 g agar in 1 L of sea water) supplemented with sterile pine needles. Petri dishes were incubated at 24 °C for one month.

Morphological structures were observed, and images captured using an optical microscope (Leica DM4500B, Leica microsystems GmbH, Wetzlar) equipped with a camera (Leica DFC320, Leica microsystems GmbH, Wetzlar). Macro- and microscopic features were compared with the available description of Roussoellaceae and Thyridariaceae [14,15,17,18,20].

2.3. DNA Extraction, PCR Amplification, and Data Assembling

Genomic DNA was extracted from about 100 mg of fresh mycelium grown on MEASW plates. Mycelium was disrupted by the mean of a MM400 tissue lyzer (Retsch GmbH, Haan, Germany) and DNA extracted using a NucleoSpin kit (Macherey Nagel GmbH, Duren, DE, USA) following the manufacturer's instructions. The quality and quantity of DNA were measured spectrophotometrically (Infinite 200 PRO NanoQuant; TECAN, Männedorf); DNA was stored at −20 °C.

The partial sequences of five genetic markers were amplified by PCR. Primer pairs ITS1/ITS4 [23], LR0R/LR7 [24], NS1/NS4 [23] were used to amplify the internal transcribed spacers, including the 5.8S rDNA gene (nrITS), 28S large ribosomal subunit (nrLSU) and 18S small ribosomal subunit (nrSSU). The translation elongation factor (TEF1α) and RNA polymerase II subunit (RPB2) were amplified by using primer pairs EF1-1018F/EF1-1620R [25] and fRPB2-5F/fPB2-7R [26].

Amplifications were run in a T100 Thermal Cycler (Bio-Rad, Hercules, CA, USA) programmed as described in Table 2. Reaction mixtures consisted of 20–40 ng DNA template, 10× PCR Buffer (15 mM $MgCl_2$, 500 mM KCl, 100 mM Tris-HCl, pH 8.3), 200 µM each dNTP, 1 µM each primer, 2.5 U Taq DNA Polymerase (Qiagen, Chatsworth, CA, USA), in 50 µL final volume. For problematic cases, additional $MgCl_2$ and/or 2.5% DMSO facilitated the reaction.

Table 2. Primers and PCR conditions used to amplify specific gene marker.

	Forward and Reverse Primers	Thermocycler Conditions	References
ITS	ITS1- ITS4	95 °C: 5 min, (95 °C: 40 s, 55 °C: 50 s, 72 °C: 50 sec) × 35 cycles; 72 °C: 8 min; 4 °C: ∞	[23]
LSU	LR0R-LR7	95 °C: 5 min, (95 °C: 1 min, 50 °C: 1 min, 72 °C: 2 min) × 35 cycles; 72 °C: 10 min; 4 °C: ∞	[24]
SSU	NS1-NS4	95 °C: 5 min, (95 °C: 1 min, 50 °C: 1 min, 72 °C: 2 min) × 35 cycles; 72 °C: 10 min; 4 °C: ∞	[23]
TEF-1α	1018F/1620R	95 °C: 5 min, (95 °C: 1 min, 50 °C: 1 min; 72 °C: 2 min) × 40 cycles, 72 °C: 10 min; 4 °C: ∞	[25]
RPB2	fRPB2-5F/fPB2-7cR	94 °C: 3 min, (94 °C: 30 s; 55 °C: 30 s; 72 °C: 1 min) × 40 cycles, 72 °C: 10 min; 4 °C: ∞	[26]

Amplicons, together with a GelPilot 1 kb plus DNA Ladder, were visualized on a 1.5% agarose gel stained with 5 mL 100 mL^{-1} ethidium bromide; PCR products were purified and sequenced at the Macrogen Europe Laboratory (Madrid, Span). The resulting Applied Biosystem (ABI) chromatograms were inspected, trimmed and assembled to obtain consensus sequences using Sequencer 5.0 (GeneCodes Corporation, Ann Arbor, Michigan, USA http://www.genecodes.com). Newly generated sequences were deposited in GenBank (Table 1).

2.4. Sequence Alignment and Phylogenetic Analysis

A dataset consisting of nrSSU, nrITS, nrLSU, TEF1α and RPB2 was assembled on the basis of BLASTn results and of recent phylogenetic studies focused on Roussoellaceae and Thyridariaceae [18,20]. Reference sequences were retrieved from GenBank (Table 1).

Sequences were aligned using MUSCLE (default conditions for gap openings and gap extension penalties), implemented in MEGA v. 7.0 (Molecular Evolutionary Genetics Analysis), visually inspected and trimmed by TrimAl v. 1.2 (http://trimal.cgenomics.org) to delimit and discard ambiguously aligned regions. Since no incongruence was observed among single-loci phylogenetic trees, alignments were concatenated into a single data matrix with SequenceMatrix [27]. The best evolutionary model under the Akaike Information Criterion (AIC) was determined with jModelTest 2 [28].

Phylogenetic inference was estimated using Maximum Likehood (ML) and Bayesian Inference (BI) criteria. The ML analysis was generated using RAxML v. 8.1.2 [29] under GTR + I + G evolutionary model and 1000 bootstrap replicates. Support values from bootstrapping runs (MLB) were mapped on the globally best tree using the "-f a" option of RAxML and "-x 12345" as a random seed to invoke the novel rapid bootstrapping algorithm. BI was performed with MrBayes 3.2.2 [30] with the same substitution model (GTR + I + G). The alignment was run for 10 million generations with two independent runs each containing four Markov Chains Monte Carlo (MCMC) and sampling every 100 iterations. The first 25% of generated trees were discarded as "burn-in". A consensus tree was generated using the "sumt" function of MrBayes and Bayesian posterior probabilities (BPP) were calculated. Consensus trees were visualized in FigTree v. 1.4.2 (http://tree.bio.ed.ac.uk/software/figtree).

Two strains of *Occutibambusa bambusae* (Occultibambusaceae) were used to root the tree. Due to topological similarity of the two resulting trees, only ML analysis with MLB and BPP values was reported (Figure 1).

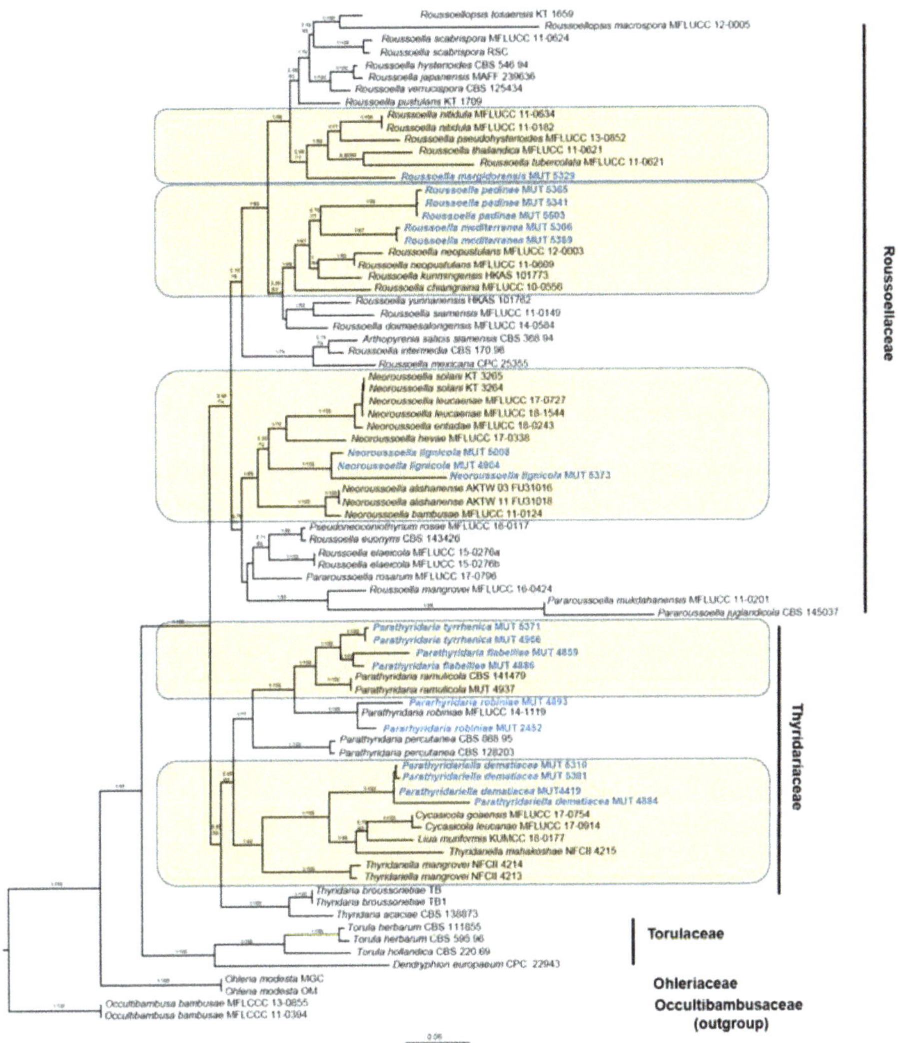

Figure 1. Phylogram generated from RAxML analysis based on a combined dataset of nrITS, nrSSU, nrLSU, TEF1α and RPB2 partial sequences. The tree is rooted to Occultibambusa bambusae. Branch numbers indicate BYPP/MLB values; Bar = expected changes per site (0.06).

DNA diagnostic characters were visually identified by the presence of heterozygous bases. For each locus, aligned sequences of the individual clusters containing new species, were inspected. Nucleotide diversities of the novel species were annotated when occurred (Tables S1–S18).

Sequence alignments and phylogenetic trees were deposited in TreeBASE (http://www.treebase.org, submission number S24773).

Following phylogenetic tree inspection, isolates that clustered in the same group and that derived from the same substrate were subjected to PCR-fingerprinting by using the micro- and mini-satellite primers $(GTG)_5$ and M13 [31,32] to exclude duplicates from further analysis. DNA fingerprints were visualized with 1.5% agarose gel stained with 5 mL 100 mL^{-1} ethidium bromide while a GelPilot 1 kb plusDNA Ladder was used as a reference. Images were acquired with a Gel Doc1000 System (Bio-Rad, Hercules, CA, USA) and fingerprints analyzed using Bionumerics v 7.6 (http://www.applied-maths.com).

3. Results

3.1. Phylogenetic Inference

Preliminary analyses carried out individually with nrITS, nrSSU, nrLSU, TEF1α and RPB2 denoted no incongruence in the topology of the single-locus trees. The combined five-markers dataset—built on the basis of BLASTn results and of recent phylogenetic studies [18,20]—consisted of 81 taxa (including MUT isolates) that represented 16 genera and 56 species (Table 1). A total of 63 sequences (2 nrITS, 8 nrSSU, 13 nrLSU, 20 TEF1α and 20 RPB2) were newly generated while 261 were retrieved from GenBank.

The combined dataset had an aligned length of 3390 characters, of which 1683 were constant, 657 were parsimony-uninformative and 1050 parsimony informative (TL = 218, CI = 0.422018, RI = 0.825243, RC = 0.348267, HI = 0.877952).

Strains MUT 4893 and MUT 2452 were identified as *Parathyridaria robiniae*, the rest of the strains represented seven new species and one new genus (Figure 1). *Parathyridaria tyrrhenica* sp. nov. (MUT 5371 and MUT 4966) formed a sister clade to *Parathyridaria flabelliae* sp. nov. (MUT 4859 and MUT 4886) with high statistical support (BYPP = 1.00; MLB = 100%); these two novel species are closely related to *P. ramulicola* (BYPP = 1.00; MLB = 100%) and clustered with other *Parathyridaria* species in the Thyridariaeae family. Within this family, four isolates (MUT 5310, MUT 5381, MUT 4419 and MUT 4884) clustered together with the genera *Thyridariella*, *Liua* and *Cycasicola*, and formed a strongly supported monophyletic lineage (BYPP = 1.00; MLB = 100%). Therefore, we have introduced the novel genus *Parathyridariella*, typified by the new species *Parathyridariella dematiacea* sp. nov.

The three strains, MUT 4904, MUT 5373 and MUT 5008, represented a novel species *Neoroussoella lignicola* sp. nov. and formed an independent and robust clade (BYPP = 1.00; MLB = 100%), within the *Neoroussoella* group in the Roussoellaceae.

Two sister clades within the *Roussoella* group were represented by the new species *Roussoella padinae* sp. nov. (MUT 5503, MUT 5341 and MUT 5365) and *Roussoella mediterranea* sp. nov. (MUT 5306 and MUT 5369). Finally, MUT 5329 *Roussoella margidorensis* sp. nov. clustered together with *R. nitidula*, *R. pseudohysterioides*, *R. thailandica* and *R. tubercolata* (BYPP = 0.99; MLB = 71%) but was phylogenetically distant from these species.

Nucleotide divergence between each novel species and members of the same clusters were annotated for each locus, when occurred (Tables S1–S18).

3.2. Taxonomy

Parathyridariella gen. nov. V. Prigione, A. Poli, E. Bovio and G.C. Varese

MYCOBANK: MB 832836

Type species. Parathyridariella dematiacea sp. nov.

Etymology. In reference to the phylogenetic proximity to the genus *Thyridariella*.

Phylogenetic placement. Thyridariaceae, Sordariomycetes, Ascomycota. The genus *Parathyridariella* gen. nov. clusters together with genera *Cycasicola*, *Liua* and *Thyridariella* (Figure 1).

Parathyridariella dematiacea sp. nov. V. Prigione, A. Poli, E. Bovio and G.C. Varese

MYCOBANK: MB 832837

Figure 2

Type. Italy, Tuscany, Mediterranean Sea, Elba Island (LI), Ghiaie ISL, 14–15 m depth, 42°49′04″N, 10°19′20″E, form the green alga *Flabellia petiolata*, 20 March 2010, R. Mussat-Sartor and N. Nurra, MUT 4884 holotype, living culture permanently preserved in metabolically inactive state by deep-freezing at Mycotheca Universitatis Taurinensis (MUT).

Additional material examined. Italy, Ligury, Mediterranean Sea, Riva Trigoso, Punta Manara (GE), 5–21 m depth, 44°15′08.62″N 9°24′17.64″E, from the seagrass *Posidonia oceanica*, March 2008, MUT 4419.

Etymology. In reference to the color of the colony on culture media.

Description. Growing actively on *Pinus pinaster* and *Quercus ruber* cork. Showing a floccose growth mainly on *Pinus pinaster*. *Hyphae* 2.8–4.8 μm wide, septate, hyaline to lightly pigmented. *Chlamydospores* numerous, mostly in chain, intercalary or solitary, globose to subglobose, from brownish to dark brown, 7–10 × 6–8 μm diameter.

Sexual morph not observed. Asexual morph with differentiated conidiogenesis not observed.

Colony description. Colonies on MEASW attaining 28–34 mm diam after 28 days at 24 °C, mycelium from dark grey/black to dark green, dense with radial grooves and concentric rings, submerged edges; reverse dark green. Brown exudate present above the concentric rings. Growth on OASW reaching 40–54 mm diam at 24 °C and 21–29 mm diam at 15 °C; colonies on PDA attaining 36–49 mm diam and 15.5–22.5 mm diam at 24 °C and 15 °C, respectively.

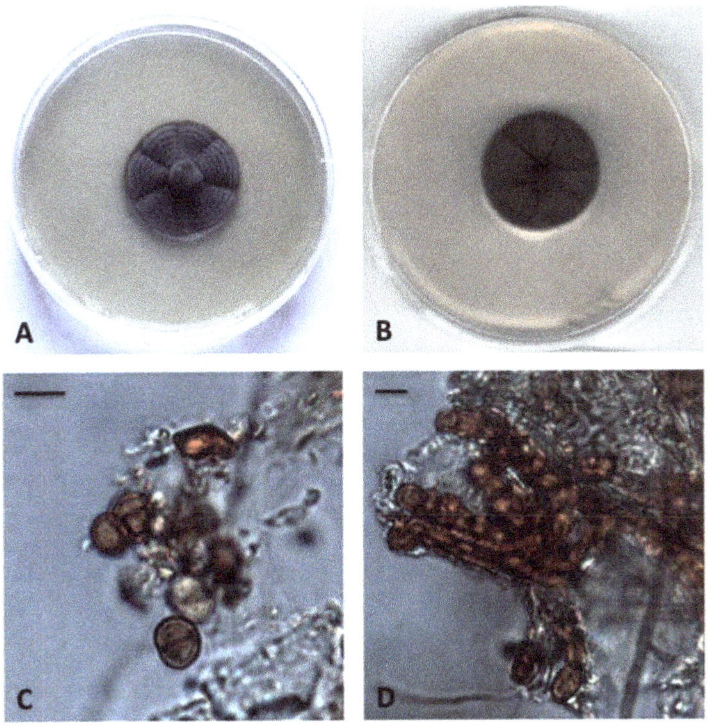

Figure 2. *Parathyridariella dematiacea* sp. nov. 28-days-old colony at 21 °C on MEASW (**A**) and reverse (**B**); solitary (**C**) and in chain (**D**) chlamydospores. Scale bars: 10 μm (C, D).

Parathyridaria tyrrhenica sp. nov. A. Poli, V. Prigione, E. Bovio and G.C. Varese
MYCOBANK: MB 832838
Figure 3

Type. Italy, Tuscany, Mediterranean Sea, Elba Island (LI), Ghiaie ISL, 14–15 m depth, 42°49′04″N, 10°19′20″E, from the brown alga *Padina pavonica*, March 2010, R. Mussat-Sartor and N. Nurra, MUT 5371 holotype, living culture permanently preserved in metabolically inactive state by deep-freezing at MUT.

Additional material examined. Italy, Tuscany, Mediterranean Sea, Elba Island (LI), Ghiaie ISL, 14–15 m depth, 42°49′04″N, 10°19′20″E, from the green alga *Flabellia petiolata,* March 2010, R. Mussat-Sartor and N. Nurra, MUT 4966.

Etymology. In reference to Tyrrhenian Sea.

Description. Growing actively on *Pinus pinaster* wood and *Quercus ruber* cork. *Hyphae* 5 μm diameter, septate, hyaline to brownish, sometimes wavy or swollen, forming hyphal strands.

Sexual morph not observed. Asexual morph with differentiated conidiogenesis: not observed.

Colony description. Colonies growing on MEASW, reaching 10 mm diam after 28 days, at 21 °C, mycelium funiculose, yellowish, lightly ochre at the edges; reverse light yellow, lighter at the edges. Growth on OASW reaching 48–50 mm diam at 24 °C and 26–29 mm diam at 15 °C; colonies on PDA attaining 31–46 mm diam and 16–19 mm diam at 24 °C and 15 °C, respectively.

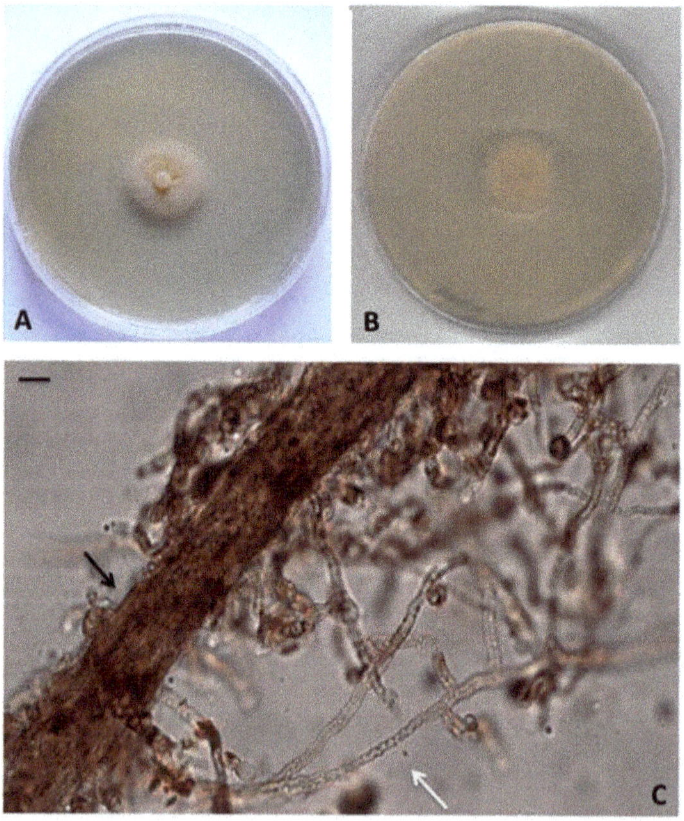

Figure 3. *Parathyridaria tyrrhenica* sp. nov. 28-days-old colony at 21 °C on MEASW (**A**) and reverse (**B**); mycelium (**C**), black and white arrows indicate hyphal strands and wavy hyphae, respectively. Scale bar: 10 μm.

Parathyridaria flabelliae sp. nov. E. Bovio, A. Poli, V. Prigione and G.C. Varese
MYCOBANK: MB 832839

Figure 4

Type. Italy, Tuscany, Mediterranean Sea, Elba Island (LI), Ghiaie ISL, 14–15 m depth, 42°49′04″N, 10°19′20″E, from the green alga *Flabellia petiolata*, March 2010, R. Mussat-Sartor and N. Nurra, MUT 4859 holotype, living culture permanently preserved in metabolically inactive state by deep-freezing at MUT.

Additional material examined. Italy, Tuscany, Mediterranean Sea, Elba Island (LI), Ghiaie ISL, 14–15 m depth, 42°49′04″N, 10°19′20″E, from the green alga *Flabellia petiolata*, March 2010, R. Mussat-Sartor and N. Nurra, MUT 4886.

Etymology. In reference to the original substratum, the green alga *Flabellia petiolata*.

Description. Growing actively on *Pinus pinaster* and on *Quercus ruber* cork. *Hyphae* 2.6–5 µm wide, septate and hyaline. *Chlamydospores* numerous, globose or subglobose, from light to dark brown, unicellular (4 × 5 µm diameter) and multicellular (up to four-celled; 8 × 12 µm diameter).

Sexual morph not observed. Asexual morph with differentiated conidiogenesis not observed.

Colony description. Colonies growing on MEASW, reaching 37–44 mm diam after 28 days at 21 °C, funiculose, whitish with submerged edges; reverse brown in the middle, lighter at edges. Growth on OASW reaching 60 mm diam at 24 °C and 33–35 mm diam at 15 °C; colonies on PDA attaining 53–64 mm diam and 23–24 mm diam at 24 °C and 15 °C, respectively.

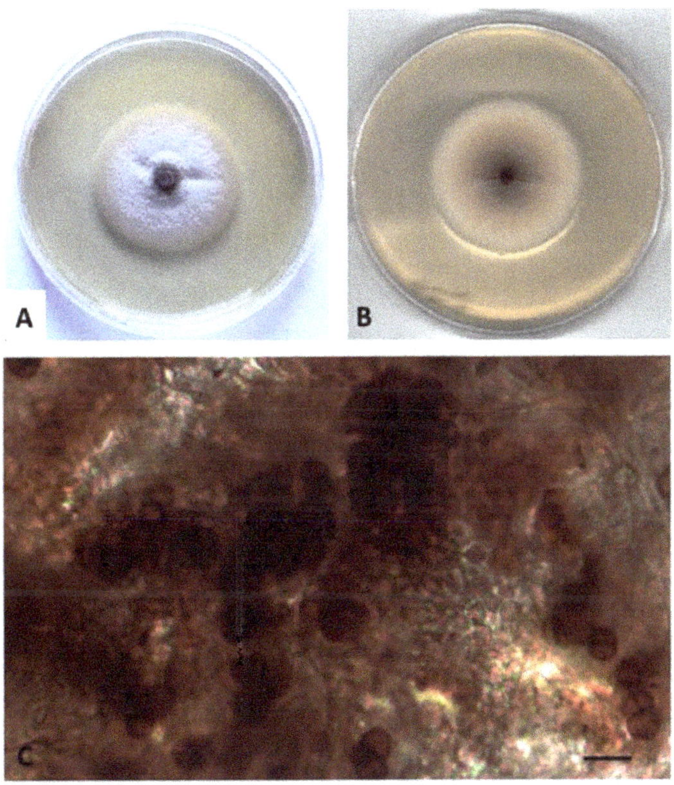

Figure 4. *Parathyridaria flabelliae* sp. nov. 28-days-old colony at 21 °C on MEASW (**A**) and reverse (**B**); unicellular and multicellular chlamydospores (**C**). Scale bar: 10 µm.

Neoroussoella lignicola sp. nov. A. Poli, E. Bovio, V. Prigione and G.C. Varese
MYCOBANK: MB 832840

Figure 5

Type. Italy, Tuscany, Mediterranean Sea, Elba Island (LI), Margidore ISL, 14–15 m depth, UTM WGS84 42°45′29″N, 10°18′24″E, from the brown alga *Padina pavonica*, March 2010, R. Mussat-Sartor and N. Nurra, MUT 5373 holotype, living culture permanently preserved in metabolically inactive state by deep-freezing at MUT.

Additional material examined. Italy, Tuscany, Mediterranean Sea, Elba Island (LI), Margidore ISL, 14–15 m depth, UTM WGS84 42°45′29″N, 10°18′24″E, from the brown alga *Padina pavonica*, March 2010, R. Mussat-Sartor and N. Nurra, MUT 4904.

Italy, Tuscany, Mediterranean Sea, Elba Island (LI), Margidore ISL, 14–15 m depth, UTM WGS84 42°45′29″N, 10°18′24″E, from the seagrass *Posidonia oceanica*, March 2010, R. Mussat-Sartor and N. Nurra, MUT 5008.

Etymology. In reference to the lignicolous behavior.

Description. Growing efficiently on *Pinus pinaster* wood. *Hyphae* 2–4.4 μm wide, septate, hyaline, assuming toruloid aspect when growing into wood vessels and forming chains of two-celled chlamydospores which, at maturity, protrude from the vessels. *Chlamydospores* 7.4 × 5.2 μm, from light to dark brown, globose or subglobose.

Sexual morph not observed. Asexual morph with differentiated conidiogenesis not observed.

Colony description. Colonies growing on MEASW, reaching 28–29 mm diam after 28 days at 21° C, from grey to dark green, floccose with irregular edges, reverse dark grey. Clear exudate often present. Growth on OASW reaching 27–40 mm diam at 24 °C and 14.5–26 mm diam at 15 °C; colonies on PDA attaining 38–45 mm diam and 19–29 mm diam at 24 °C and 15 °C, respectively.

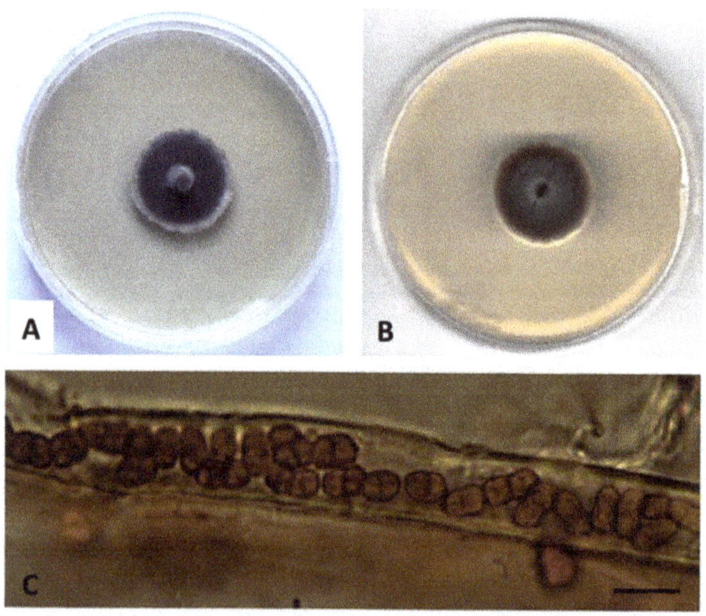

Figure 5. *Neoroussoella lignicola* sp. nov. 28-days-old colony at 21 °C on MEASW (**A**) and reverse (**B**); two-celled chlamydospores inside wood vessels (**C**). Scale bar: 10 μm.

Roussoella margidorensis sp. nov. E. Bovio, V. Prigione, A. Poli and G.C. Varese
MYCOBANK: MB 832841
Figure 6

Type. Italy, Tuscany, Mediterranean Sea, Elba Island (LI), Margidore ISL, 14–15 m depth, UTM WGS84 42°45′29″N, 10°18′24″E, from the brown alga *Padina pavonica*, March 2010, R. Mussat-Sartor and N. Nurra, MUT 5329 holotype, living culture permanently preserved in metabolically inactive state by deep-freezing at MUT.

Etymology. In reference to the area of origin, Margidore.

Description. Growing actively on *Pinus pinaster* wood. *Hyphae* approx. 2 μm wide, septate, brownish.

Sexual morph not observed. Asexual morph and differentiated conidiogenesis not observed.

Colony description. Colonies growing on MEASW, attaining 33–34 mm diam after 28 days at 21 °C; whitish, lighter to the edge, umbonate in the middle, reverse ochre. Caramel diffusible pigment produced. Growth on OASW reaching 45 mm diam at 24 °C and 27 mm diam at 15 °C; colonies on PDA attaining 45 mm diam and 23 mm diam at 24 °C and 15 °C, respectively.

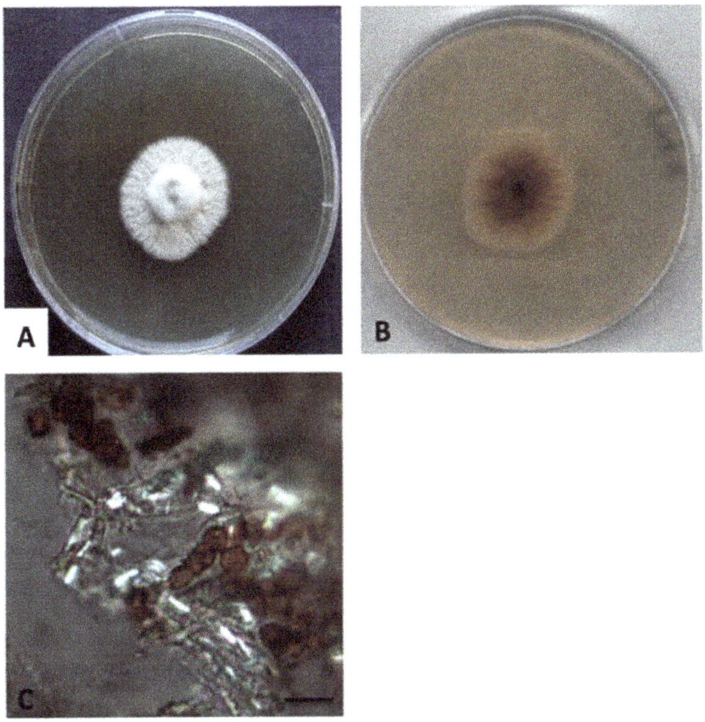

Figure 6. *Roussoella margidorensis* sp. nov. 28-days-old colony at 21 °C on MEASW (**A**) and reverse (**B**); chlamydospores (**C**). Scale bar: 10 μm.

Roussoella mediterranea sp. nov. A. Poli, E. Bovio, V. Prigione, and G.C. Varese
MYCOBANK: MB 832842
Figure 7

Type. Italy, Tuscany, Mediterranean Sea, Elba Island (LI), Margidore ISL, 14–15 m depth, UTM WGS84 42°45′29″N, 10°18′24″E, from the brown alga *Padina pavonica*, March 2010, R. Mussat-Sartor and N. Nurra, MUT 5369 holotype, living culture permanently preserved in metabolically inactive state by deep-freezing at MUT.

Additional material examined. Italy, Tuscany, Mediterranean Sea, Elba Island (LI), Margidore ISL, 14–15 m depth, UTM WGS84 42°45′29″N, 10°18′24″E, from the brown alga *Padina pavonica*, March

2010, R. Mussat-Sartor and N. Nurra, MUT 5306 (identical to MUT 5306 on the basis of micro- and minisatellite analyses)

Etymology. In reference to the geographical origin, Mediterranean Sea.

Description in culture. Growing actively on *Pinus pinaster* wood and poorly colonizing *Quercus ruber* cork. *Hyphae* 2.4 μm wide, septate, dematiaceous. *Chlamydospores* 4.5 × 5.7 μm, from unicellular to 4-celled; branched chains of light to dark brown chlamydospores often present.

Sexual morph not observed. Asexual morph with differentiated conidiogenesis not observed.

Colony description. Colonies growing on MEASW, reaching 55 mm diam after 28 days at 21 °C, light grey, floccose, with umbonate area in the middle, reverse brown with lighter edges. Dark exudate present. Growth on OASW reaching 67–72 mm diam at 24 °C and 33–38 mm diam at 15 °C; colonies on PDA attaining 69–76 mm diam and 32.5–39 mm diam at 24 °C and 15 °C, respectively.

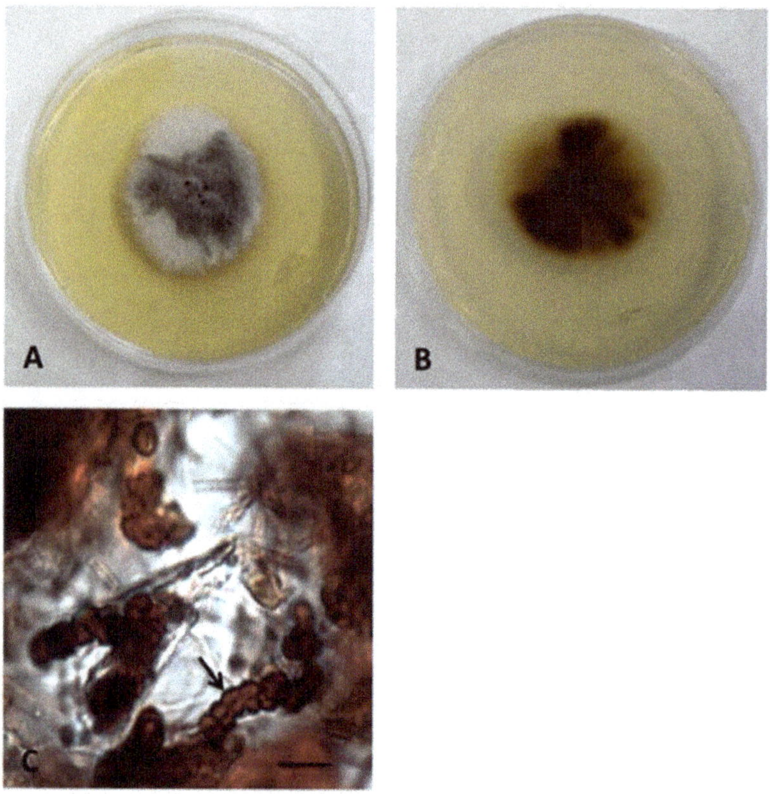

Figure 7. *Roussoella mediterranea* sp. nov. 28-days-old colony at 21 °C on MEASW (**A**) and reverse (**B**); unicellular and multicellular chlamydosporesn indicated by a black arrow (**C**). Scale bar: 10 μm.

Roussoella padinae sp. nov. V. Prigione, E. Bovio, A. Poli and G.C. Varese
MYCOBANK: MB 832843
Figure 8

Type. Italy, Tuscany, Mediterranean Sea, Elba Island (LI), Margidore ISL, 14–15 m depth, UTM WGS84 42°45′29″N, 10°18′24″E, from the brown alga *Padina pavonica*, March 2010, R. Mussat-Sartor and N. Nurra, MUT 5503 holotype, living culture permanently preserved in metabolically inactive state by deep-freezing at MUT.

Additional material examined. Italy, Tuscany, Mediterranean Sea, Elba Island (LI), Margidore ISL, 14–15 m depth, UTM WGS84 42°45′29″N, 10°18′24″E, from the brown alga *Padina pavonica*, March 2010, R. Mussat-Sartor and N. Nurra, MUT 5341 and MUT 5345 (identical to MUT 5503 on the basis of micro- and minisatellite analyses)

Etymology. In reference to the original substratum, *Padina pavonica*.

Description in culture. Growing efficiently on *Quercus ruber* cork and poorly colonizing *Pinus pinaster* wood. *Hyphae* 3 µm wide, septate, brownish, assuming toruloid aspect when growing into wood vessels and forming chains of two-celled chlamydospores which, at maturity, protrude from the vessels. *Chlamydospores* 5–7 × 4 µm, from light to dark brown, subglobose, ellipsoidal or cylindrical.

Sexual morph not observed. Asexual morph n with differentiated conidiogenesis not observed.

Colony description. Colonies growing on MEASW, reaching 53 mm diam after 28 days at 21 °C, from grey to dark green, floccose in the middle, with radial grooves, fimbriate edges, reverse brown. Growth on OASW reaching 57.5–65 mm diam at 24 °C and 30–35 mm diam at 15 °C; colonies on PDA attaining 60–69 mm diam and 30–34 mm diam at 24 °C and 15 °C, respectively.

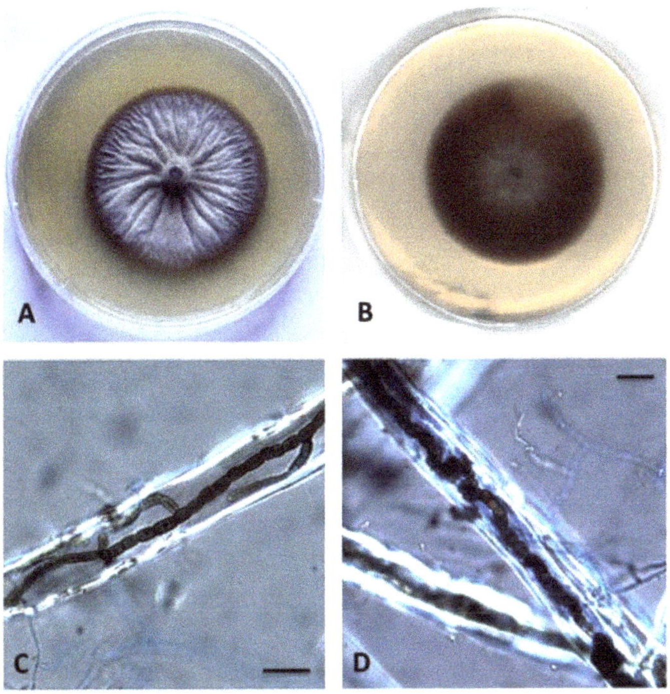

Figure 8. *Roussoella padinae* sp. nov. 28-days-old colony at 21 °C on MEASW (**A**) and reverse (**B**); toruloid hyphae (**C**) and two-celled chlamydospores (**D**) inside wood vessels. Scale bars: 10 µm.

4. Discussion

The description of these new taxa was particularly challenging because neither asexual nor sexual reproductive structures developed in axenic conditions. Therefore, we were unable to describe the range of anatomical variations and diagnostic features among these newly recognized phylogenetic lineages. Indeed, strictly vegetative growth without sporulation is a common feature of many marine fungal strains [10,11,33]. Possibly, these organisms rely on hyphal fragmentation for their dispersal, or alternatively, the differentiation of reproductive structures may be obligatorily dependent on the peculiar environmental conditions under which they live (e.g., wet-dry cycles, high salinity, low temperature, high pressure, etc.). During the study of these fungi, we tried to mimic the saline environment by using different culture media supplemented with natural sea water or sea salts. Although these culture methods were applied to induce sporulation, we observed that only media supplemented with sea water supported a measurable growth of vegetative mycelium (data not shown). The method introduced by Panebianco et al. [22] to induce sporulation by placing wood and cork specimens on the colony surface with their subsequent transfer into sea water, was only partially successful: out of seven species, three (*P. dematiacea, P. flabelliae, R. mediterranea*) developed chlamydospores in the mycelium above the wood surface, two (*N. lignicola, R. padinae*) gave rise to resting spores inside wood vessels. Most of the strains preferred to colonize *P. pinaster* wood rather than *Q. ruber* cork. These structures were interpreted as "chlamydospores" instead of "conidia" for the following reasons: (i) They were characterized by a very thick cell wall, a typical feature of resting spores; (ii) conidiogenous cells were never observed. Additional efforts to force the development of reproductive structures by using SNASW and pine needles, were also unsuccessful.

Both *R. padinae* and *N. lignicola* displayed a similar lignicolous behavior, growing and producing chlamydospores inside wooden vessels, although of different size and shape. The ability to form hyphae and to grow inside the wood vessels has been reported for a number of dark septate endophyte fungi in terrestrial environment [34] and, recently, for *Posidoniomyces atricolor* Vohník and Réblová, a marine endophyte that lives in association with the roots of *P. oceanica* [35]. By definition, endophytes live inside living plant tissues. To induce sporulation, sterilized specimens of dead wood were employed, therefore *R. padinae* and *N. lignicola* were inferred to be "lignicolous fungi" rather than "endophytes". The observation of this growth characteristic in two different genera, may find its reason in an evolutionary adaptation to marine life in association with lignocellulosic matrices. Therefore, we can hypothesize their ecological role as saprobes involved in degrading organic matter.

Notwithstanding the lack of exhaustive descriptions of morphological features, the strongly supported phylogenetic and molecular analysis, conducted with five different genetic markers (nrSSU, nrITS, nrLSU, TEF1α and RPB2) undoubtedly pointed out the differences among these species and their belonging to new taxa. This is also supported by the DNA diagnostic characters identified in the individual loci (Tables S1–S18). In particular, the present study introduces four new species of Roussoellaceae and three new species of Thyridariaceae. Indeed, only MUT 2452 and MUT 4893 were ascribable to the previously described *P. robiniae* (Figure 1). In the case of MUT 4884, the holotype of *P. dematiacea*, a novel genus was proposed since it formed a defined cluster with MUT 5310 and MUT 4419, well separated by the genera *Cycasicola*, *Liua* and *Thyridariella*.

Most of the Roussoellaceae and Thyridariaceae described to date are associated with terrestrial plants, especially bamboo and palm species [15,16]. In fact, only two species, *R. mangrovei* and *R. nitidula* have been retrieved from the marine environment (www.marinefungi.org). However, considering the present study, we can infer that these families are well represented in the sea, thus improving our knowledge on the largely unexplored fungal marine biodiversity.

Supplementary Materials: The following are available online at http://www.mdpi.com/1424-2818/12/4/144/s1, Table S1: The eight variable sites detected in the nrITS region among *P. dematiacea* and its neighbor species, Table S2: The single variable site detected in the nrLSU region among *P. dematiacea* and its neighbor species, Table S3: The five variable sites detected in the nrSSU region among *P. dematiacea* and its neighbor species, Table S4: The six variable sites detected in the TEF1α partial gene among *P. dematiacea* and its neighbor species, Table S5: The six variable sites detected in the nrITS region among *P. tyrrhenica*, *P. flabelliae* and their neighbor species, Table S6: The eight variable sites detected in the nrLSU region among *P. tyrrhenica*, *P. flabelliae* and their neighbor species, Table S7: The eight variable sites detected in the TEF1α partial gene among *P. tyrrhenica*, *P. flabelliae* and their neighbor species, Table S8: The 33 variable sites detected in the RPB2 partial gene among *P. tyrrhenica*, *P. flabelliae* and their neighbor species, Table S9: The two variable sites detected in nrITS region among *R. mediterranea*, *R. padinae* and the neighbor species, Table S10: The single variable site detected in nrLSU region among *R. mediterranea*, *R. padinae*, and the neighbor species, Table S11: The six sites detected in the TEF1α partial gene among *R. mediterranea*, *R. padinae* and the neighbor species, Table S12: The six sites detected in the RPB2 partial gene among *R. mediterranea*, *R. padinae* and the neighbor species, Table S13: The eight variable sites detected in the nrITS region among *N. lignicola* and its neighbor species, Table S14: The three variable sites detected in the nrLSU region among *N. lignicola* and its neighbor species, Table S15: The eight variable sites detected in the nrSSU region among *N. lignicola* and its neighbor species, Table S16: The ten sites detected in the TEF1α partial gene among *N. lignicola* and its neighbor species, Table S17: The three variable sites detected in the nrITS region among *R. margidoriensis* and its neighbor species, Table S18: The 29 variable sites detected in the TEF1α partial gene among *R. margidoriensis* and its neighbor species

Author Contributions: Conceptualization, A.P., E.B., V.P., G.C.V.; methodology, A.P., E.B., V.P., G.C.V.; software, A.P.; validation, A.P., E.B., V.P., L.R., G.C.V.; formal analysis, A.P., E.B., V.P., L.R.; investigation, A.P., E.B., V.P., L.R., G.C.V.; resources, V.P., G.C.V.; data curation, A.P., V.P.; writing—original draft preparation, A.P., V.P.; writing—review and editing, A.P., E.B., V.P., G.C.V.; visualization, A.P., V.P.; supervision, V.P., G.C.V.; project administration, A.P., E.B., V.P., G.C.V.; funding acquisition, G.C.V. All authors have read and agreed to the published version of the manuscript.

Funding: This research was funded by Fondazione CRT, Torino, Italy and by the University of Torino (ex 60%).

Acknowledgments: The authors are grateful to Pelagosphera s.c.r.l. for harvesting algal and seagrass samples.

Conflicts of Interest: The authors declare no conflict of interest.

References

1. Richards, T.A.; Jones, M.D.; Leonard, G.; Bass, D. Marine fungi: Their ecology and molecular diversity. *Ann. Rev. Mar. Sci.* **2012**, *4*, 495–522. [CrossRef] [PubMed]
2. Amend, A.; Burgaud, G.; Cunliffe, M.; Edgcomb, V.P.; Ettinger, C.L.; Gutierrez, M.H.; Heitman, J.; Hom, E.F.Y.; Ianiri, G.; Jones, A.C.; et al. Fungi in the marine environment: Open questions and unsolved problems. *Mbio* **2019**, *10*, 15. [CrossRef] [PubMed]
3. Carroll, A.R.; Copp, B.R.; Davis, R.A.; Keyzers, R.A.; Prinsep, M.R. Marine natural products. *Nat. Prod. Rep.* **2019**, *36*, 122–173. [CrossRef] [PubMed]
4. Jones, E.G.; Pang, K.-L. *Marine fungi and fungal-like organisms*; Jones, E.G., Pang, K.-L., Eds.; Walter de Gruyter: Berlin, Germany, 2012.
5. Raghukumar, S. *Fungi in Coastal and Oceanic Marine Ecosystems: Marine Fungi*; Springer: Cham, Switzerland, 2017.
6. Pang, K.L.; Overy, D.P.; Jones, E.B.G.; Calado, M.D.; Burgaud, G.; Walker, A.K.; Johnson, J.A.; Kerr, R.G.; Cha, H.J.; Bills, G.F. 'Marine fungi' and 'marine-derived fungi' in natural product chemistry research: Toward a new consensual definition. *Fungal Biol. Rev.* **2016**, *30*, 163–175. [CrossRef]
7. Jones, E.B.G.; Pang, K.-L.; Abdel-Wahab, M.A.; Scholz, B.; Hyde, K.D.; Boekhout, T.; Ebel, R.; Rateb, M.E.; Henderson, L.; Sakayaroj, J.; et al. An online resource for marine fungi. *Fungal Divers.* **2019**, *96*, 347–433. [CrossRef]
8. Jones, E.B.G. Fifty years of marine mycology. *Fungal Divers.* **2011**, *50*, 73–112. [CrossRef]
9. Jones, E.B.G.; Suetrong, S.; Sakayaroj, J.; Bahkali, A.H.; Abdel-Wahab, M.A.; Boekhout, T.; Pang, K.L. Classification of marine Ascomycota, Basidiomycota, Blastocladiomycota and Chytridiomycota. *Fungal Divers.* **2015**, *73*, 1–72. [CrossRef]
10. Garzoli, L.; Poli, A.; Prigione, V.; Gnavi, G.; Varese, G.C. Peacock's tail with a fungal cocktail: First assessment of the mycobiota associated with the brown alga *Padina pavonica*. *Fungal Ecol.* **2018**, *35*, 87–97. [CrossRef]
11. Gnavi, G.; Garzoli, L.; Polil, A.; Prigione, V.; Burgaud, G.; Varese, G.C. The culturable mycobiota of *Flabellia petiolata*: First survey of marine fungi associated to a Mediterranean green alga. *PLoS ONE* **2017**, *12*. [CrossRef]
12. Panno, L.; Bruno, M.; Voyron, S.; Anastasi, A.; Gnavi, G.; Miserere, L.; Varese, G.C. Diversity, ecological role and potential biotechnological applications of marine fungi associated to the seagrass *Posidonia oceanica*. *New Biotechnol.* **2013**, *30*, 685–694. [CrossRef]
13. Bovio, E.; Garzoli, L.; Poli, A.; Prigione, V.; Firsova, D.; McCormack, G.; Varese, G. The culturable mycobiota associated with three Atlantic sponges, including two new species: *Thelebolus balaustiformis* and *T. spongiae*. *Fungal Syst. Evol.* **2018**, *1*, 141–167. [CrossRef]
14. Liu, J.-K.; Phookamsak, R.; Dai, D.-Q.; Tanaka, K.; Jones, E.; Xu, J.-C.; Chukeatirote, E.; Hyde, K.D. Roussoellaceae, a new pleosporalean family to accommodate the genera *Neoroussoella* gen. nov., *Roussoella* and *Roussoellopsis*. *Phytotaxa* **2014**, *181*, 1–33. [CrossRef]
15. Jaklitsch, W.M.; Voglmayr, H. Hidden diversity in *Thyridaria* and a new circumscription of the Thyridariaceae. *Stud. Mycol.* **2016**, *85*, 35–64. [CrossRef]
16. Tibpromma, S.; Hyde, K.D.; Jeewon, R.; Maharachchikumbura, S.S.N.; Liu, J.K.; Bhat, D.J.; Jones, E.B.G.; McKenzie, E.H.C.; Camporesi, E.; Bulgakov, T.S.; et al. Fungal diversity notes 491–602: Taxonomic and phylogenetic contributions to fungal taxa. *Fungal Divers.* **2017**, *83*, 1–261. [CrossRef]

17. Devadatha, B.; Sarma, V.V.; Jeewon, R.; Wanasinghe, D.N.; Hyde, K.D.; Jones, E.B.G. *Thyridariella*, a novel marine fungal genus from India: Morphological characterization and phylogeny inferred from multigene DNA sequence analyses. *Mycol. Prog.* **2018**, *17*, 791–804. [CrossRef]
18. Jayasiri, S.; Hyde, K.; Jones, E.; McKenzie, E.; Jeewon, R.; Phillips, A.; Bhat, D.; Wanasinghe, D.; Liu, J.; Lu, Y.; et al. Diversity, morphology and molecular phylogeny of Dothideomycetes on decaying wild seed pods and fruits. *Mycosphere* **2019**, *10*, 1–186. [CrossRef]
19. Phookamsak, R.; Hyde, K.D.; Jeewon, R.; Bhat, D.J.; Jones, E.B.G.; Maharachchikumbura, S.S.N.; Raspe, O.; Karunarathna, S.C.; Wanasinghe, D.N.; Hongsanan, S.; et al. Fungal diversity notes 929–1035: Taxonomic and phylogenetic contributions on genera and species of fungi. *Fungal Divers.* **2019**, *95*, 1–273. [CrossRef]
20. Jiang, H.B.; Hyde, K.D.; Jayawardena, R.S.; Doilom, M.; Xu, J.C.; Phookamsak, R. Taxonomic and phylogenetic characterizations reveal two new species and two new records of *Roussoella* (Roussoellaceae, Pleosporales) from Yunnan, China. *Mycol. Prog.* **2019**, *18*, 577–591. [CrossRef]
21. Dai, D.Q.; Phookamsak, R.; Wijayawardene, N.N.; Li, W.J.; Bhat, D.J.; Xu, J.C.; Taylor, J.E.; Hyde, K.D.; Chukeatirote, E. Bambusicolous fungi. *Fungal Divers.* **2017**, *82*, 1–105. [CrossRef]
22. Panebianco, C.; Tam, W.Y.; Jones, E.B.G. The effect of pre-inoculation of balsa wood by selected marine fungi and their effect on subsequent colonisation in the sea. *Fungal Divers.* **2002**, *10*, 77–88.
23. White, T.J.; Bruns, T.; Lee, S.; Taylor, J. Amplification and direct sequencing of fungal ribosomal RNA genes for phylogenetics. In *PCR protocols: A guide to methods and applications*; Academic Press: New York, NY, USA, 1990; Volume 18, pp. 315–322.
24. Vilgalys, R.; Hester, M. Rapid genetic identification and mapping of enzymatically amplified ribosomal DNA from several *Cryptococcus* species. *J. Bacteriol.* **1990**, *172*, 4238–4246. [CrossRef]
25. Stielow, J.; Lévesque, C.; Seifert, K.; Meyer, W.; Iriny, L.; Smits, D.; Renfurm, R.; Verkley, G.; Groenewald, M.; Chaduli, D. One fungus, which genes? Development and assessment of universal primers for potential secondary fungal DNA barcodes. *Persoonia* **2015**, *35*, 242–263. [CrossRef]
26. Liu, Y.J.; Whelen, S.; Hall, B.D. Phylogenetic relationships among ascomycetes: Evidence from an RNA polymerse II subunit. *Mol. Biol. Evol.* **1999**, *16*, 1799–1808. [CrossRef]
27. Vaidya, G.; Lohman, D.J.; Meier, R. SequenceMatrix: Concatenation software for the fast assembly of multi-gene datasets with character set and codon information. *Cladistics* **2011**, *27*, 171–180. [CrossRef]
28. Darriba, D.; Taboada, G.L.; Doallo, R.; Posada, D. jModelTest 2: More models, new heuristics and parallel computing. *Nat. Methods* **2012**, *9*, 772. [CrossRef]
29. Stamatakis, A. RAxML version 8: A tool for phylogenetic analysis and post-analysis of large phylogenies. *Bioinformatics* **2014**, *30*, 1312–1313. [CrossRef]
30. Ronquist, F.; Teslenko, M.; van der Mark, P.; Ayres, D.L.; Darling, A.; Hohna, S.; Larget, B.; Liu, L.; Suchard, M.A.; Huelsenbeck, J.P. MrBayes 3.2: Efficient Bayesian Phylogenetic Inference and Model Choice Across a Large Model Space. *Syst. Biol.* **2012**, *61*, 539–542. [CrossRef]
31. Lieckfeldt, E.; Meyer, W.; Borner, T. Rapid identification and differentiation of yeasts by DNA and PCR fingerprinting. *J. Basic Microbiol.* **1993**, *33*, 413–426. [CrossRef]
32. Poli, A.; Lazzari, A.; Prigione, V.; Voyron, S.; Spadaro, D.; Varese, G.C. Influence of plant genotype on the cultivable fungi associated to tomato rhizosphere and roots in different soils. *Fungal Biol.* **2016**, *120*, 862–872. [CrossRef]
33. Poli, A.; Vizzini, A.; Prigione, V.; Varese, G.C. Basidiomycota isolated from the Mediterranean Sea-Phylogeny and putative ecological roles. *Fungal Ecol.* **2018**, *36*, 51–62. [CrossRef]
34. Oses, R.; Valenzuela, S.; Freer, J.; Sanfuentes, E.; Rodriguez, J. Fungal endophytes in xylem of healthy Chilean trees and their possible role in early wood decay. *Fungal Divers.* **2008**, *33*, 77–86.
35. Vohnik, M.; Borovce, O.; Kolarikova, Z.; Sudova, R.; Reblova, M. Extensive sampling and high-throughput sequencing reveal *Posidoniomyces atricolor* gen. et sp. nov. (Aigialaceae, Pleosporales) as the dominant root mycobiont of the dominant Mediterranean seagrass Posidoniaoceanica. *Mycokeys* **2019**, *55*, 59–86. [CrossRef] [PubMed]

MDPI
St. Alban-Anlage 66
4052 Basel
Switzerland
Tel. +41 61 683 77 34
Fax +41 61 302 89 18
www.mdpi.com

Diversity Editorial Office
E-mail: diversity@mdpi.com
www.mdpi.com/journal/diversity